STUDENTS STRUCTURAL HANDBOOK

With Revision Notes and Formulae

Students Structural Handbook

With Revision Notes and Formulae

W. MORGAN
B. Sc., A.M.I. Struct. E.

Revised by

I. HAMILTON
D.I.C., C. Eng., M.I.C.E.

Principal
Kenya Highways Training Centre
Nairobi

LONDON

BUTTERWORTHS

THE BUTTERWORTH GROUP

England:
Butterworth & Co (Publishers) Ltd
London: 88 Kingsway, WC2B 6AB

Australia:
Butterworths Pty Ltd
Sydney: 586 Pacific Highway, NSW 2067
Melbourne: 343 Little Collins Street, 3000
Brisbane: 240 Queen Street, 4000

Canada:
Butterworth & Co (Canada) Ltd
Toronto: 14 Curity Avenue, 374

New Zealand:
Butterworths of New Zealand Ltd
Wellington: 26-28 Waring Taylor Street, 1

South Africa:
Butterworth & Co (South Africa) (Pty) Ltd
Durban: 152-154 Gale Street

First published in 1967
Second edition 1973

ISBN 0 408 70361 X Standard
0 408 70362 8 Limp

PRINTED BY Unwin Brothers Limited
THE GRESHAM PRESS OLD WOKING SURREY ENGLAND

Produced by 'Uneoprint'

A member of the Staples Printing Group (G3747)

Preface

This handbook is written mainly for the use of Higher National Certificate and architectural students. There are many excellent textbooks dealing with theory of structures and also with design in particular materials. Those who take structures as a main subject will refer to the appropriate specialised manuals. For students who need a more general knowledge of structural topics, this book gives a guide to Codes of Practice and simple design methods in the main building materials. Some topics cannot be suitably simplified for description in a book of this kind and in most of those cases a reference has been given.

Rewriting the book in SI units has presented problems as usage has not yet standardised. Certain Codes and Standards were not available in SI units and some metric equivalents have been given. It is hoped that publication at this still uncertain stage will help those who are studying and lecturing on structural topics with very few textbooks as yet issued in SI units.

I. HAMILTON

Contents

Section 1

General

SI Units, Conversion Factors, SI Notation
Mathematical Tables
Loading
Bending Moments, Deflections and Section
Properties
Definitions

1. General

1.1 SI UNITS

The SI (Système International d'Unités) is based on six units (Table 1.1). From the point of view of the structural student, the most important derived units are those shown in Table 1.1.2. The biggest innovation in SI is the introduction of a different unit, the newton, for force, from that used for mass, the kilogram. The newton is that force which, applied to a mass of 1 kilogram, gives it an acceleration of 1 metre/second^2.

Table 1.1

Quantity	Unit	Symbol
Length	Metre	m
Mass	Kilogram	kg
Time	Second	s
Electric current	Ampere	A
Temperature	Kelvin	K
Luminous intensity	Candela	cd

Table 1.1.2 Some derived SI units

Quantity	Unit	Symbol
Force	Newton	N
Area	Square metre	m^2
	Square millimetre	mm^2
Volume	Cubic metre	m^3
	Cubic millimetre	mm^3
Length	Millimetre	mm
Density	Kilogram per cubic metre	kg/m^3
Pressure and Stress	Newton per square metre	N/m^2
	Newton per square millimetre	N/mm^2
	Kilonewton per square metre	kN/m^2
Mass	Tonne ($=10^3$ kg)	t

Table 1.1.3 Multiples of SI units

Multiplication factor		Prefix	Symbol
1 000 000	10^6	mega	M
1000	10^3	kilo	k
100	10^2	hecto	h
10	10	deca	da
0·1	10^{-1}	deci	d
0·01	10^{-2}	centi	c
0·001	10^{-3}	milli	m
0·000 001	10^{-6}	micro	μ

CONVERSION FACTORS

Available conversion tables include SI Conversion Charts (Butterworths) and the Architect's Journal Metric Handbook. A few conversion factors are given below.

SI unit	Imperial unit	SI to Imperial	Imperial to SI
Metre (m)		3·281 ft	
	Foot (ft)		0·305 m
Millimetre (mm)		0·0394 in	
	Inch (in)		25·4 mm
Kilogram (kg)		2·205 lb	
	Pound (lb)		0·454 kg
Newton (N)		0·225 lbf	
	Pound-force (lbf)		4·448 N
7 newtons per square millimetre		1000 lbf/in^2	
15 newtons per square millimetre		1 tonf/in^2	

SI notation

The decimal marker is the conventional decimal point, raised from the line, not the comma used in some early British publications and still in use in some metric countries. The full stop on the line is used in typewriting unless the machine has a special character above the line. Values less than unity should always have a zero before the decimal point, so that we write 0·600, for example, not ·600. No full stop is used for abbreviations used as symbols for units, so that the abbreviation for millimetre is mm not mm., for example.

When there is a group of five or more digits to left or right of the decimal point, they are grouped in threes with a space, not a comma, dividing them, thus:

0·1041

0·104 11

0·104 111

1041·0

11 041·0

111 041·0

On drawings, dimensions are always in metres or millimetres, indicated by figures only, with no symbol. A dimension in metres always has three digits after the decimal point.

Thus:	1·740 0·320 14·000	all metres
	14 112·5 2741	all millimetres

Table 1.2.1

LOGARITHMS

	0	1	2	3	4	5	6	7	8	9	Mean Differences 1	2	3	4	5	6	7	8	9
10	0000	0043	0086	0128	0170	0212	0253	0294	0334	0374	4	8	12	17	21	25	29	33	37
11	0414	0453	0492	0531	0569	0607	0645	0682	0719	0755	4	8	11	15	19	23	26	30	34
12	0792	0828	0864	0899	0934	0969	1004	1038	1072	1106	3	7	10	14	17	21	24	28	31
13	1139	1173	1206	1239	1271	1303	1335	1367	1399	1430	3	6	10	13	16	19	23	26	29
14	1461	1492	1523	1553	1584	1614	1644	1673	1703	1732	3	6	9	12	15	18	21	24	27
15	1761	1790	1818	1847	1875	1903	1931	1959	1987	2014	3	6	8	11	14	17	20	22	25
16	2041	2068	2095	2122	2148	2175	2201	2227	2253	2279	3	5	8	11	13	16	18	21	24
17	2304	2330	2355	2380	2405	2430	2455	2480	2504	2529	2	5	7	10	12	15	17	20	22
18	2553	2577	2601	2625	2648	2672	2695	2718	2742	2765	2	5	7	9	12	14	16	19	21
19	2788	2810	2833	2856	2878	2900	2923	2945	2967	2989	2	4	7	9	11	13	16	18	20
20	3010	3032	3054	3075	3096	3118	3139	3160	3181	3201	2	4	6	8	11	13	15	17	19
21	3222	3243	3263	3284	3304	3324	3345	3365	3385	3404	2	4	6	8	10	12	14	16	18
22	3424	3444	3464	3483	3502	3522	3541	3560	3579	3598	2	4	6	8	10	12	14	15	17
23	3617	3636	3655	3674	3692	3711	3729	3747	3766	3784	2	4	6	7	9	11	13	15	17
24	3802	3820	3838	3856	3874	3892	3909	3927	3945	3962	2	4	5	7	9	11	12	14	16
25	3979	3997	4014	4031	4048	4065	4082	4099	4116	4133	2	3	5	7	9	10	12	14	15
26	4150	4166	4183	4200	4216	4232	4249	4265	4281	4298	2	3	5	7	8	10	11	13	15
27	4314	4330	4346	4362	4378	4393	4409	4425	4440	4456	2	3	5	6	8	9	11	13	14
28	4472	4487	4502	4518	4533	4548	4564	4579	4594	4609	2	3	5	6	8	9	11	12	14
29	4624	4639	4654	4669	4683	4698	4713	4728	4742	4757	1	3	4	6	7	9	10	12	13
30	4771	4786	4800	4814	4829	4843	4857	4871	4886	4900	1	3	4	6	7	9	10	11	13
31	4914	4928	4942	4955	4969	4983	4997	5011	5024	5038	1	3	4	6	7	8	10	11	12
32	5051	5065	5079	5092	5105	5119	5132	5145	5159	5172	1	3	4	5	7	8	9	11	12
33	5185	5198	5211	5224	5237	5250	5263	5276	5289	5302	1	3	4	5	6	8	9	10	12
34	5315	5328	5340	5353	5366	5378	5391	5403	5416	5428	1	3	4	5	6	8	9	10	11
35	5441	5453	5465	5478	5490	5502	5514	5527	5539	5551	1	2	4	5	6	7	9	10	11
36	5563	5575	5587	5599	5611	5623	5635	5647	5658	5670	1	2	4	5	6	7	8	10	11
37	5682	5694	5705	5717	5729	5740	5752	5763	5775	5786	1	2	3	5	6	7	8	9	10
38	5798	5809	5821	5832	5843	5855	5866	5877	5888	5899	1	2	3	5	6	7	8	9	10
39	5911	5922	5933	5944	5955	5966	5977	5988	5999	6010	1	2	3	4	5	7	8	9	10
40	6021	6031	6042	6053	6064	6075	6085	6096	6107	6117	1	2	3	4	5	6	8	9	10
41	6128	6138	6149	6160	6170	6180	6191	6201	6212	6222	1	2	3	4	5	6	7	8	9
42	6232	6243	6253	6263	6274	6284	6294	6304	6314	6325	1	2	3	4	5	6	7	8	9
43	6335	6345	6355	6365	6375	6385	6395	6405	6415	6425	1	2	3	4	5	6	7	8	9
44	6435	6444	6454	6464	6474	6484	6493	6503	6513	6522	1	2	3	4	5	6	7	8	9
45	6532	6542	6551	6561	6571	6580	6590	6599	6609	6618	1	2	3	4	5	6	7	8	9
46	6628	6637	6646	6656	6665	6675	6684	6693	6702	6712	1	2	3	4	5	6	7	7	8
47	6721	6730	6739	6749	6758	6767	6776	6785	6794	6803	1	2	3	4	5	5	6	7	8
48	6812	6821	6830	6839	6848	6857	6866	6875	6884	6893	1	2	3	4	4	5	6	7	8
49	6902	6911	6920	6928	6937	6946	6955	6964	6972	6981	1	2	3	4	4	5	6	7	8
50	6990	6998	7007	7016	7024	7033	7042	7050	7059	7067	1	2	3	3	4	5	6	7	8
51	7076	7084	7093	7101	7110	7118	7126	7135	7143	7152	1	2	3	3	4	5	6	7	8
52	7160	7168	7177	7185	7193	7202	7210	7218	7226	7235	1	2	2	3	4	5	6	7	7
53	7243	7251	7259	7267	7275	7284	7292	7300	7308	7316	1	2	2	3	4	5	6	6	7
54	7324	7332	7340	7348	7356	7364	7372	7380	7388	7396	1	2	2	3	4	5	6	6	7

Table 1.2.1 (continued)

LOGARITHMS

	0	1	2	3	4	5	6	7	8	9	Mean Differences								
											1	2	3	4	5	6	7	8	9
55	7404	7412	7419	7427	7435	7443	7451	7459	7466	7474	1	2	2	3	4	5	5	6	7
56	7482	7490	7497	7505	7513	7520	7528	7536	7543	7551	1	2	2	3	4	5	5	6	7
57	7559	7566	7574	7582	7589	7597	7604	7612	7619	7627	1	2	2	3	4	5	5	6	7
58	7634	7642	7649	7657	7664	7672	7679	7686	7694	7701	1	1	2	3	4	4	5	6	7
59	7709	7716	7723	7731	7738	7745	7752	7760	7767	7774	1	1	2	3	4	4	5	6	7
60	7782	7789	7796	7803	7810	7818	7825	7832	7839	7846	1	1	2	3	4	4	5	6	6
61	7853	7860	7868	7875	7882	7889	7896	7903	7910	7917	1	1	2	3	4	4	5	6	6
62	7924	7931	7938	7945	7952	7959	7966	7973	7980	7987	1	1	2	3	3	4	5	6	6
63	7993	8000	8007	8014	8021	8028	8035	8041	8048	8055	1	1	2	3	3	4	5	5	6
64	8062	8069	8075	8082	8089	8096	8102	8109	8116	8122	1	1	2	3	3	4	5	5	6
65	8129	8136	8142	8149	8156	8162	8169	8176	8182	8189	1	1	2	3	3	4	5	5	6
66	8195	8202	8209	8215	8222	8228	8235	8241	8248	8254	1	1	2	3	3	4	5	5	6
67	8261	8267	8274	8280	8287	8293	8299	8306	8312	8319	1	1	2	3	3	4	5	5	6
68	8325	8331	8338	8344	8351	8357	8363	8370	8376	8382	1	1	2	3	3	4	4	5	6
69	8388	8395	8401	8407	8414	8420	8426	8432	8439	8445	1	1	2	2	3	4	4	5	6
70	8451	8457	8463	8470	8476	8482	8488	8494	8500	8506	1	1	2	2	3	4	4	5	6
71	8513	8519	8525	8531	8537	8543	8549	8555	8561	8567	1	1	2	2	3	4	4	5	5
72	8573	8579	8585	8591	8597	8603	8609	8615	8621	8627	1	1	2	2	3	4	4	5	5
73	8633	8639	8645	8651	8657	8663	8669	8675	8681	8686	1	1	2	2	3	4	4	5	5
74	8692	8698	8704	8710	8716	8722	8727	8733	8739	8745	1	1	2	2	3	4	4	5	5
75	8751	8756	8762	8768	8774	8779	8785	8791	8797	8802	1	1	2	2	3	3	4	5	5
76	8808	8814	8820	8825	8831	8837	8842	8848	8854	8859	1	1	2	2	3	3	4	5	5
77	8865	8871	8876	8882	8887	8893	8899	8904	8910	8915	1	1	2	2	3	3	4	4	5
78	8921	8927	8932	8938	8943	8949	8954	8960	8965	8971	1	1	2	2	3	3	4	4	5
79	8976	8982	8987	8993	8998	9004	9009	9015	9020	9025	1	1	2	2	3	3	4	4	5
80	9031	9036	9042	9047	9053	9058	9063	9069	9074	9079	1	1	2	2	3	3	4	4	5
81	9085	9090	9096	9101	9106	9112	9117	9122	9128	9133	1	1	2	2	3	3	4	4	5
82	9138	9143	9149	9154	9159	9165	9170	9175	9180	9186	1	1	2	2	3	3	4	4	5
83	9191	9196	9201	9206	9212	9217	9222	9227	9232	9238	1	1	2	2	3	3	4	4	5
84	9243	9248	9253	9258	9263	9269	9274	9279	9284	9289	1	1	2	2	3	3	4	4	5
85	9294	9299	9304	9309	9315	9320	9325	9330	9335	9340	1	1	2	2	3	3	4	4	5
86	9345	9350	9355	9360	9365	9370	9375	9380	9385	9390	1	1	2	2	3	3	4	4	5
87	9395	9400	9405	9410	9415	9420	9425	9430	9435	9440	0	1	1	2	2	3	3	4	4
88	9445	9450	9455	9460	9465	9469	9474	9479	9484	9489	0	1	1	2	2	3	3	4	4
89	9494	9499	9504	9509	9513	9518	9523	9528	9533	9538	0	1	1	2	2	3	3	4	4
90	9542	9547	9552	9557	9562	9566	9571	9576	9581	9586	0	1	1	2	2	3	3	4	4
91	9590	9595	9600	9605	9609	9614	9619	9624	9628	9633	0	1	1	2	2	3	3	4	4
92	9638	9643	9647	9652	9657	9661	9666	9671	9675	9680	0	1	1	2	2	3	3	4	4
93	9685	9689	9694	9699	9703	9708	9713	9717	9722	9727	0	1	1	2	2	3	3	4	4
94	9731	9736	9741	9745	9750	9754	9759	9763	9768	9773	0	1	1	2	2	3	3	4	4
95	9777	9782	9786	9791	9795	9800	9805	9809	9814	9818	0	1	1	2	2	3	3	4	4
96	9823	9827	9832	9836	9841	9845	9850	9854	9859	9863	0	1	1	2	2	3	3	4	4
97	9868	9872	9877	9881	9886	9890	9894	9899	9903	9908	0	1	1	2	2	3	3	4	4
98	9912	9917	9921	9926	9930	9934	9939	9943	9948	9952	0	1	1	2	2	3	3	4	4
99	9956	9961	9965	9969	9974	9978	9983	9987	9991	9996	0	1	1	2	2	3	3	3	4

Table 1.2.2

ANTILOGARITHMS

	0	1	2	3	4	5	6	7	8	9	Mean Differences 1	2	3	4	5	6	7	8	9
·00	1000	1002	1005	1007	1009	1012	1014	1016	1019	1021	0	0	1	1	1	1	2	2	2
·01	1023	1026	1028	1030	1033	1035	1038	1040	1042	1045	0	0	1	1	1	1	2	2	2
·02	1047	1050	1052	1054	1057	1059	1062	1064	1067	1069	0	0	1	1	1	1	2	2	2
·03	1072	1074	1076	1079	1081	1084	1086	1089	1091	1094	0	0	1	1	1	1	2	2	2
·04	1096	1099	1102	1104	1107	1109	1112	1114	1117	1119	0	1	1	1	1	2	2	2	2
·05	1122	1125	1127	1130	1132	1135	1138	1140	1143	1146	0	1	1	1	1	2	2	2	2
·06	1148	1151	1153	1156	1159	1161	1164	1167	1169	1172	0	1	1	1	1	2	2	2	2
·07	1175	1178	1180	1183	1186	1189	1191	1194	1197	1199	0	1	1	1	1	2	2	2	2
·08	1202	1205	1208	1211	1213	1216	1219	1222	1225	1227	0	1	1	1	1	2	2	2	3
·09	1230	1233	1236	1239	1242	1245	1247	1250	1253	1256	0	1	1	1	1	2	2	2	3
·10	1259	1262	1265	1268	1271	1274	1276	1279	1282	1285	0	1	1	1	1	2	2	2	3
·11	1288	1291	1294	1297	1300	1303	1306	1309	1312	1315	0	1	1	1	2	2	2	2	3
·12	1318	1321	1324	1327	1330	1334	1337	1340	1343	1346	0	1	1	1	2	2	2	2	3
·13	1349	1352	1355	1358	1361	1365	1368	1371	1374	1377	0	1	1	1	2	2	2	3	3
·14	1380	1384	1387	1390	1393	1396	1400	1403	1406	1409	0	1	1	1	2	2	2	3	3
·15	1413	1416	1419	1422	1426	1429	1432	1435	1439	1442	0	1	1	1	2	2	2	3	3
·16	1445	1449	1452	1455	1459	1462	1466	1469	1472	1476	0	1	1	1	2	2	2	3	3
·17	1479	1483	1486	1489	1493	1496	1500	1503	1507	1510	0	1	1	1	2	2	2	3	3
·18	1514	1517	1521	1524	1528	1531	1535	1538	1542	1545	0	1	1	1	2	2	2	3	3
·19	1549	1552	1556	1560	1563	1567	1570	1574	1578	1581	0	1	1	1	2	2	3	3	3
·20	1585	1589	1592	1596	1600	1603	1607	1611	1614	1618	0	1	1	1	2	2	3	3	3
·21	1622	1626	1629	1633	1637	1641	1644	1648	1652	1656	0	1	1	2	2	2	3	3	3
·22	1660	1663	1667	1671	1675	1679	1683	1687	1690	1694	0	1	1	2	2	2	3	3	3
·23	1698	1702	1706	1710	1714	1718	1722	1726	1730	1734	0	1	1	2	2	2	3	3	4
·24	1738	1742	1746	1750	1754	1758	1762	1766	1770	1774	0	1	1	2	2	2	3	3	4
·25	1778	1782	1786	1791	1795	1799	1803	1807	1811	1816	0	1	1	2	2	2	3	3	4
·26	1820	1824	1828	1832	1837	1841	1845	1849	1854	1858	0	1	1	2	2	3	3	3	4
·27	1862	1866	1871	1875	1879	1884	1888	1892	1897	1901	0	1	1	2	2	3	3	3	4
·28	1905	1910	1914	1919	1923	1928	1932	1936	1941	1945	0	1	1	2	2	3	3	4	4
·29	1950	1954	1959	1963	1968	1972	1977	1982	1986	1991	0	1	1	2	2	3	3	4	4
·30	1995	2000	2004	2009	2014	2018	2023	2028	2032	2037	0	1	1	2	2	3	3	4	4
·31	2042	2046	2051	2056	2061	2065	2070	2075	2080	2084	0	1	1	2	2	3	3	4	4
·32	2089	2094	2099	2104	2109	2113	2118	2123	2128	2133	0	1	1	2	2	3	3	4	4
·33	2138	2143	2148	2153	2158	2163	2168	2173	2178	2183	0	1	1	2	2	3	3	4	4
·34	2188	2193	2198	2203	2208	2213	2218	2223	2228	2234	1	1	2	2	3	3	4	4	5
·35	2239	2244	2249	2254	2259	2265	2270	2275	2280	2286	1	1	2	2	3	3	4	4	5
·36	2291	2296	2301	2307	2312	2317	2323	2328	2333	2339	1	1	2	2	3	3	4	4	5
·37	2344	2350	2355	2360	2366	2371	2377	2382	2388	2393	1	1	2	2	3	3	4	4	5
·38	2399	2404	2410	2415	2421	2427	2432	2438	2443	2449	1	1	2	2	3	3	4	4	5
·39	2455	2460	2466	2472	2477	2483	2489	2495	2500	2506	1	1	2	2	3	3	4	5	5
·40	2512	2518	2523	2529	2535	2541	2547	2553	2559	2564	1	1	2	2	3	4	4	5	5
·41	2570	2576	2582	2588	2594	2600	2606	2612	2618	2624	1	1	2	2	3	4	4	5	5
·42	2630	2636	2642	2649	2655	2661	2667	2673	2679	2685	1	1	2	2	3	4	4	5	6
·43	2692	2698	2704	2710	2716	2723	2729	2735	2742	2748	1	1	2	3	3	4	4	5	6
·44	2754	2761	2767	2773	2780	2786	2793	2799	2805	2812	1	1	2	3	3	4	4	5	6
·45	2818	2825	2831	2838	2844	2851	2858	2864	2871	2877	1	1	2	3	3	4	5	5	6
·46	2884	2891	2897	2904	2911	2917	2924	2931	2938	2944	1	1	2	3	3	4	5	5	6
·47	2951	2958	2965	2972	2979	2985	2992	2999	3006	3013	1	1	2	3	3	4	5	5	6
·48	3020	3027	3034	3041	3048	3055	3062	3069	3076	3083	1	1	2	3	4	4	5	6	6
·49	3090	3097	3105	3112	3119	3126	3133	3141	3148	3155	1	1	2	3	4	4	5	6	6

Table 1.2.2 (continued)

ANTILOGARITHMS

	0	1	2	3	4	5	6	7	8	9	Mean Differences								
											1	2	3	4	5	6	7	8	9
·50	3162	3170	3177	3184	3192	3199	3206	3214	3221	3228	1	1	2	3	4	4	5	6	7
·51	3236	3243	3251	3258	3266	3273	3281	3289	3296	3304	1	2	2	3	4	5	5	6	7
·52	3311	3319	3327	3334	3342	3350	3357	3365	3373	3381	1	2	2	3	4	5	5	6	7
·53	3388	3396	3404	3412	3420	3428	3436	3443	3451	3459	1	2	2	3	4	5	6	6	7
·54	3467	3475	3483	3491	3499	3508	3516	3524	3532	3540	1	2	2	3	4	5	6	6	7
·55	3548	3556	3565	3573	3581	3589	3597	3606	3614	3622	1	2	2	3	4	5	6	7	7
·56	3631	3639	3648	3656	3664	3673	3681	3690	3698	3707	1	2	3	3	4	5	6	7	8
·57	3715	3724	3733	3741	3750	3758	3767	3776	3784	3793	1	2	3	3	4	5	6	7	8
·58	3802	3811	3819	3828	3837	3846	3855	3864	3873	3882	1	2	3	4	4	5	6	7	8
·59	3890	3899	3908	3917	3926	3936	3945	3954	3963	3972	1	2	3	4	5	5	6	7	8
·60	3981	3990	3999	4009	4018	4027	4036	4046	4055	4064	1	2	3	4	5	6	6	7	8
·61	4074	4083	4093	4102	4111	4121	4130	4140	4150	4159	1	2	3	4	5	6	7	8	9
·62	4169	4178	4188	4198	4207	4217	4227	4236	4246	4256	1	2	3	4	5	6	7	8	9
·63	4266	4276	4285	4295	4305	4315	4325	4335	4345	4355	1	2	3	4	5	6	7	8	9
·64	4365	4375	4385	4395	4406	4416	4426	4436	4446	4457	1	2	3	4	5	6	7	8	9
·65	4467	4477	4487	4498	4508	4519	4529	4539	4550	4560	1	2	3	4	5	6	7	8	9
·66	4571	4581	4592	4603	4613	4624	4634	4645	4656	4667	1	2	3	4	5	6	7	9	10
·67	4677	4688	4699	4710	4721	4732	4742	4753	4764	4775	1	2	3	4	5	7	8	9	10
·68	4786	4797	4808	4819	4831	4842	4853	4864	4875	4887	1	2	3	4	6	7	8	9	10
·69	4898	4909	4920	4932	4943	4955	4966	4977	4989	5000	1	2	3	5	6	7	8	9	10
·70	5012	5023	5035	5047	5058	5070	5082	5093	5105	5117	1	2	4	5	6	7	8	9	11
·71	5129	5140	5152	5164	5176	5188	5200	5212	5224	5236	1	2	4	5	6	7	8	10	11
·72	5248	5260	5272	5284	5297	5309	5321	5333	5346	5358	1	2	4	5	6	7	9	10	11
·73	5370	5383	5395	5408	5420	5433	5445	5458	5470	5483	1	3	4	5	6	8	9	10	11
·74	5495	5508	5521	5534	5546	5559	5572	5585	5598	5610	1	3	4	5	6	8	9	10	12
·75	5623	5636	5649	5662	5675	5689	5702	5715	5728	5741	1	3	4	5	7	8	9	10	12
·76	5754	5768	5781	5794	5808	5821	5834	5848	5861	5875	1	3	4	5	7	8	9	11	12
·77	5888	5902	5916	5929	5943	5957	5970	5984	5998	6012	1	3	4	5	7	8	10	11	12
·78	6026	6039	6053	6067	6081	6095	6109	6124	6138	6152	1	3	4	6	7	8	10	11	13
·79	6166	6180	6194	6209	6223	6237	6252	6266	6281	6295	1	3	4	6	7	9	10	11	13
·80	6310	6324	6339	6353	6368	6383	6397	6412	6427	6442	1	3	4	6	7	9	10	12	13
·81	6457	6471	6486	6501	6516	6531	6546	6561	6577	6592	2	3	5	6	8	9	11	12	14
·82	6607	6622	6637	6653	6668	6683	6699	6714	6730	6745	2	3	5	6	8	9	11	12	14
·83	6761	6776	6792	6808	6823	6839	6855	6871	6887	6902	2	3	5	6	8	9	11	13	14
·84	6918	6934	6950	6966	6982	6998	7015	7031	7047	7063	2	3	5	6	8	10	11	13	15
·85	7079	7096	7112	7129	7145	7161	7178	7194	7211	7228	2	3	5	7	8	10	12	13	15
·86	7244	7261	7278	7295	7311	7328	7345	7362	7379	7396	2	3	5	7	8	10	12	13	15
·87	7413	7430	7447	7464	7482	7499	7516	7534	7551	7568	2	3	5	7	9	10	12	14	16
·88	7586	7603	7621	7638	7656	7674	7691	7709	7727	7745	2	4	5	7	9	11	12	14	16
·89	7762	7780	7798	7816	7834	7852	7870	7889	7907	7925	2	4	5	7	9	11	13	14	16
·90	7943	7962	7980	7998	8017	8035	8054	8072	8091	8110	2	4	6	7	9	11	13	15	17
·91	8128	8147	8166	8185	8204	8222	8241	8260	8279	8299	2	4	6	8	9	11	13	15	17
·92	8318	8337	8356	8375	8395	8414	8433	8453	8472	8492	2	4	6	8	10	12	14	15	17
·93	8511	8531	8551	8570	8590	8610	8630	8650	8670	8690	2	4	6	8	10	12	14	16	18
·94	8710	8730	8750	8770	8790	8810	8831	8851	8872	8892	2	4	6	8	10	12	14	16	18
·95	8913	8933	8954	8974	8995	9016	9036	9057	9078	9099	2	4	6	8	10	12	15	17	19
·96	9120	9141	9162	9183	9204	9226	9247	9268	9290	9311	2	4	6	8	11	13	15	17	19
·97	9333	9354	9376	9397	9419	9441	9462	9484	9506	9528	2	4	7	9	11	13	15	17	20
·98	9550	9572	9594	9616	9638	9661	9683	9705	9727	9750	2	4	7	9	11	13	16	18	20
·99	9772	9795	9817	9840	9863	9886	9908	9931	9954	9977	2	5	7	9	11	14	16	18	20

Table 1.2.3 Trigonometrical ratios

Angle							
Degrees	Radians	Sine	Tangent	Co-tangent	Cosine		
0°	0	0	0	∞	1	1·5708	90°
1	·0175	·0175	·0175	57·2900	·9998	1·5533	89
2	·0349	·0349	·0349	28·6363	·9994	1·5359	88
3	·0524	·0523	·0524	19·0811	·9986	1·5184	87
4	·0698	·0698	·0698	14·3007	·9976	1·5010	86
5	·0873	·0872	·0875	11·4301	·9962	1·4835	85
6	·1047	·1045	·1051	9·5144	·9945	1·4661	84
7	·1222	·1219	·1228	8·1443	·9925	1·4486	83
8	·1396	·1392	·1405	7·1154	·9903	1·4312	82
9	·1571	·1564	·1584	6·3138	·9877	1·4137	81
10	·1745	·1736	·1763	5·6713	·9848	1·3963	80
11	·1920	·1908	·1944	5·1446	·9816	1·3788	79
12	·2094	·2079	·2126	4·7046	·9781	1·3614	78
13	·2269	·2250	·2309	4·3315	·9744	1·3439	77
14	·2443	·2419	·2493	4·0108	·9703	1·3265	76
15	·2618	·2588	·2679	3·7321	·9659	1·3090	75
16	·2793	·2756	·2867	3·4874	·9613	1·2915	74
17	·2967	·2924	·3057	3·2709	·9563	1·2741	73
18	·3142	·3090	·3249	3·0777	·9511	1·2566	72
19	·3316	·3256	·3443	2·9042	·9455	1·2392	71
20	·3491	·3420	·3640	2·7475	·9397	1·2217	70
21	·3665	·3584	·3839	2·6051	·9336	1·2043	69
22	·3840	·3746	·4040	2·4751	·9272	1·1868	68
23	·4014	·3907	·4245	2·3559	·9205	1·1694	67
24	·4189	·4067	·4452	2·2460	·9135	1·1519	66
25	·4363	·4226	·4663	2·1445	·9063	1·1345	65
26	·4538	·4384	·4877	2·0503	·8988	1·1170	64
27	·4712	·4540	·5095	1·9626	·8910	1·0996	63
28	·4887	·4695	·5317	1·8807	·8829	1·0821	62
29	·5061	·4848	·5543	1·8040	·8746	1·0647	61
30	·5236	·5000	·5774	1·7321	·8660	1·0472	60
31	·5411	·5150	·6009	1·6643	·8572	1·0297	59
32	·5585	·5299	·6249	1·6003	·8480	1·0123	58
33	·5760	·5446	·6494	1·5399	·8387	·9948	57
34	·5934	·5592	·6745	1·4826	·8290	·9774	56
35	·6109	·5736	·7002	1·4281	·8192	·9599	55
36	·6283	·5878	·7265	1·3764	·8090	·9425	54
37	·6458	·6018	·7536	1·3270	·7986	·9250	53
38	·6632	·6157	·7813	1·2799	·7880	·9076	52
39	·6807	·6293	·8098	1·2349	·7771	·8901	51
40	·6981	·6428	·8391	1·1918	·7660	·8727	50
41	·7156	·6561	·8693	1·1504	·7547	·8552	49
42	·7330	·6691	·9004	1·1106	·7431	·8378	48
43	·7505	·6820	·9325	1·0724	·7314	·8203	47
44	·7679	·6947	·9657	1·0355	·7193	·8029	46
45°	·7854	·7071	1·0000	1·0000	·7071	·7854	45°
		Cosine	Co-tangent	Tangent	Sine	Radians	Degrees
						Angle	

Table 1. 3. 1 Minimum imposed floor loads

Type of building or structure	Distributed load		Concentrated load over a 300 mm square	
	kN/m²	kgf/m²	kN	kgf
Assembly buildings—with fixed seating	4·0	408	—	—
—without fixed seating	5·0	510	3·6	367
Bedrooms — domestic buildings	1·5	153	1·4	143
— hotels	2·0	204	1·8	184
— institutions	1·5	153	1·8	184
Book stores	2·4 for each metre of storage	245	7·0	714
Clubs — assembly areas with fixed seating	4·0	408	—	—
assembly areas without fixed seating	5·0	510	3·6	367
Colleges — classrooms	3·0	306	2·7	275
— dining rooms	2·0	204	2·7	275
— dormitories	1·5	153	1·8	184
— gymnasia	5·0	510	3·6	367
Corridors etc and footbridges between buildings:				
— subject to crowd loadings	4·0	408	4·5	459
— subject to loads greater than crowds	5·0	510	4·5	459
Dance halls	5·0	510	3·6	367
Factories and similar buildings	5·0	510	4·5	459
	7·5	765	6·7	714
	or 10·0	or 1020	9·0	918
Garages — car parking for vehicles not exceeding 2500 kg	2·5	255	9·0	918
— all repair workshops and parking for vehicles exceeding 2500 kg	5·0	510	9·0	918
Hospitals — wards	2·0	204	1·8	184
utility rooms, operating theatres	2·0	204	4·5	459
Hotels — bars and vestibules	5·0	510	—	—
kitchens	3·0	306	4·5	459
Houses	1·5	153	1·4	143
Offices — filing and storage	5·0	510	4·5	459
general offices	2·5	255	2·7	275
offices with computing equipment etc.	3·5	357	4·5	459
Stairs — houses less than 3 stories	1·5	153	1·8	184
all other buildings—the same as the floors to which they give access	3·0 to 5·0	306 to 510		
Stationery stores	4·0 for each metre of storage height	408	9·0	918

1.3 LOADING

CP 3: Chapter V: Part 1: 1967, deals with dead and imposed loads. All floor slabs are required to carry the appropriate distributed or concentrated imposed loads, whichever produces the greater stresses. Beams are to be designed to carry the distributed load appropriate to their use. If no values are given for concentrated load, it may be assumed that the distributed load is adequate for design purposes (see Table 1.3.1).

Table 1.3.2 Reduction in total distributed imposed floor loads

Number of floors, including roof carried by member	1	2	3	4	5–10	10+
Percentage reduction in load on all floors carried by member	0	10	20	30	40	50

If a single span of a beam carries more than 46 m^2 of floor at one general level the imposed load can be reduced by 5 per cent for each 46 m^2, subject to a maximum reduction of 25 per cent.

Flat roofs and sloping roofs up to 10°

If access in addition to that needed for cleaning and repair is to be provided the imposed load is 1·5 kN/m^2 (153 kgf/m^2) or a load of 1·8 kN (184 kgf) concentrated on a 300 mm square.

If no access is provided other than that required for cleaning and repair, the imposed load is 0·75 kN/m^2 (76·5 kgf/m^2) or a load of 0·9 kN (91·8 kgf) concentrated on a 300 mm square.

Sloping roofs

If the slope is greater than 10°, with no access provided other than that needed for cleaning and repair, the following imposed loads apply:

1. Up to 30° a load of 0·75 kN/m^2 (76·5 kgf/m^2) measured on plan or a vertical load of 0·9 kN (91·8 kgf) on a square with a 300 mm side.
2. No imposed load is to be considered if the slope is greater greater than 75°. For slopes from 30° to 75° the imposed load can be found by linear interpolation.

Table 1. 3. 3 Horizontal loads on parapets and balustrades

Use	Intensity of horizontal load N/m run	kgf/m run
Light access stairs etc less than 600 mm wide	220	22·4
Light access stairs etc more than 600 mm wide to private and domestic premises	360	36·7
All other stairways etc, parapets to roofs	740	75·5
Panic barriers	3000	306

Wind loading

CP 3: Chapter V: Part 2: 1967 deals with wind loads. The treatment of wind loading is very much more complicated than in previous codes. It is not possible to summarise the requirements adequately, and reference must be made to the code itself for wind loading.

Table 1.3.4 Mass densities of materials

Material	kg/m³
Aluminium	2771
Asbestos cement	1922-2082
Asphalt	2082
Bitumen roofing felt	593
Brass	8426
Brickwork, commons	2000
heavy pressed brick	2240
engineering	2400
Cement	1441
Concrete, plain	2300
reinforced	2400
Copper	8730
Cork	128-240
Felt, roofing	593
Fibre building board	160-400
Floors — hollow-clay blocks with concrete ribs between blocks and 40 mm concrete topping	1600
Glass, plate	2787
Lead	11 325
Plaster, acoustic	800
fibrous	480
gypsum	1920
Steel, mild	7849
Stone, limestone	2082-2243
sandstone	2195-2403
granite	2595-2931
Timber, oak	721-961
pitchpine	673
douglas fir	529

1.4 BENDING MOMENTS, DEFLECTIONS, AND SECTION PROPERTIES

Table 1.4.1 Maximum bending moments and maximum deflections

Loading	Maximum Bending Moment	Values of C Maximum deflection $= \frac{CWL^3}{EI}$
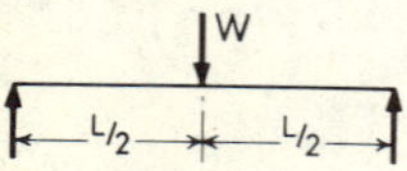 (a)	$\frac{WL}{4}$	$\frac{1}{48}$
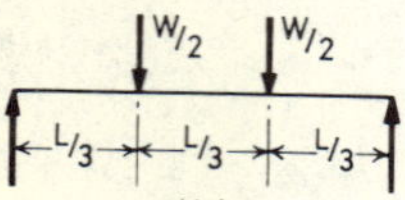 (b)	$\frac{WL}{6}$	$\frac{23}{1296}$
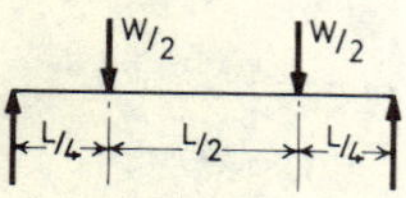 (c)	$\frac{WL}{8}$	$\frac{11}{768}$
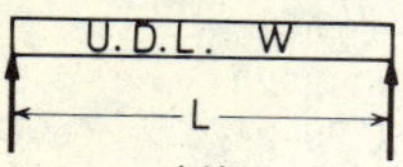 (d)	$\frac{WL}{8}$	$\frac{5}{384}$

Table 1.4.1 (continued)

Loading	Maximum Bending Moment	Values of C Maximum deflection $= \frac{CWL^3}{EI}$
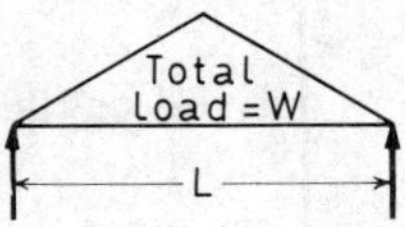 (e)	$\frac{WL}{6}$	$\frac{1}{60}$
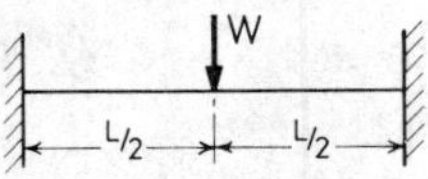 (f)	$\frac{WL}{8}$ (at supports and at mid-span)	$\frac{1}{192}$
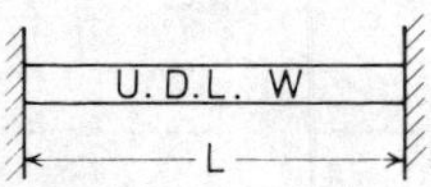 (g)	$\frac{WL}{12}$ at supports $\frac{WL}{24}$ at mid-span	$\frac{1}{384}$
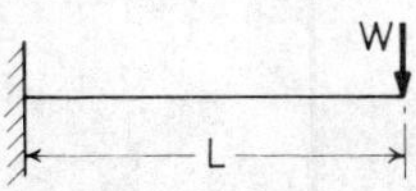 (h)	WL	$\frac{1}{3}$
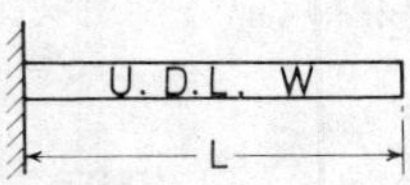 (i)	$\frac{WL}{2}$	$\frac{1}{8}$

For bending moments in reinforced beams and slabs spanning in one direction, see Table 3.1.1.

Table 1.4.2 Properties of sections

Symbols:

A = area of cross-section (mm^2)

y = distance from neutral axis to extreme fibres (mm) (It is this distance which is used in the formula $M = f(I/y)$ when designing beams.)

XX and YY are axes which pass through the centroid of the section

I_{XX} and I_{YY} = moments of inertia (i.e. second moments of area) about the axes XX and YY (mm^4)

Z_{XX} and Z_{YY} = section moduli about the axes XX and YY (mm^3)

r_{XX} and r_{YY} = radii of gyration about axes XX and YY (mm)

$$r = \sqrt{\frac{I}{A}}$$

Rectangle

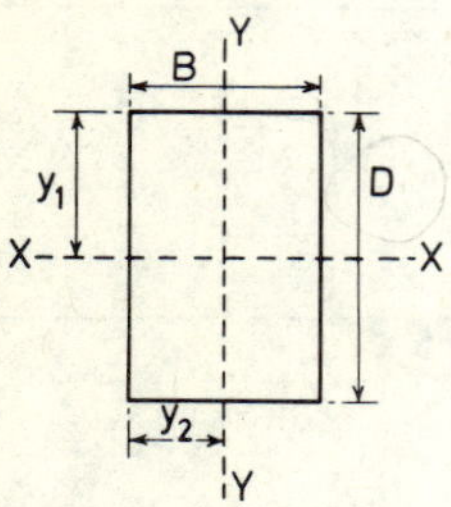

(a)

$A = BD$

$$y_1 = \frac{D}{2}, \quad y_2 = \frac{B}{2}$$

$$I_{XX} = \frac{BD^3}{12} \quad I_{YY} = \frac{DB^3}{12}$$

$$Z_{XX} = \frac{BD^2}{6} \quad Z_{YY} = \frac{DB^2}{6}$$

$$r_{XX} = \sqrt{\frac{D}{12}} = 0{\cdot}289D, \; r_{YY} = 0{\cdot}289B$$

Table 1.4.2 (continued)

Hollow Rectangle

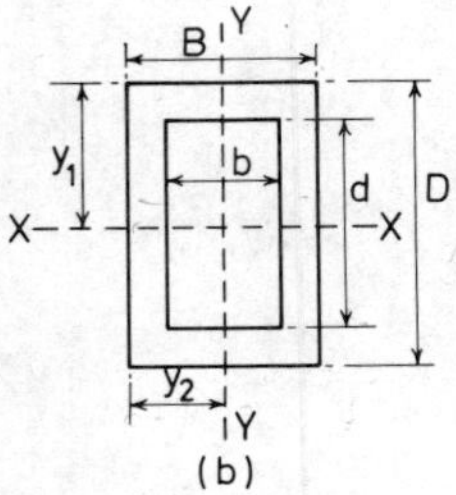

$$A = BD - bd$$

$$y_1 = \frac{D}{2} \qquad y_2 = \frac{B}{2}$$

$$I_{XX} = \frac{BD^3 - bd^3}{12} \qquad I_{YY} = \frac{DB^3 - db^3}{12}$$

$$Z_{XX} = \frac{BD^3 - bd^3}{6D} \qquad Z_{YY} = \frac{DB^3 - db^3}{6B}$$

$$r_{XX} = \sqrt{\frac{I_{XX}}{A}} \qquad r_{YY} = \sqrt{\frac{I_{YY}}{A}}$$

I-section

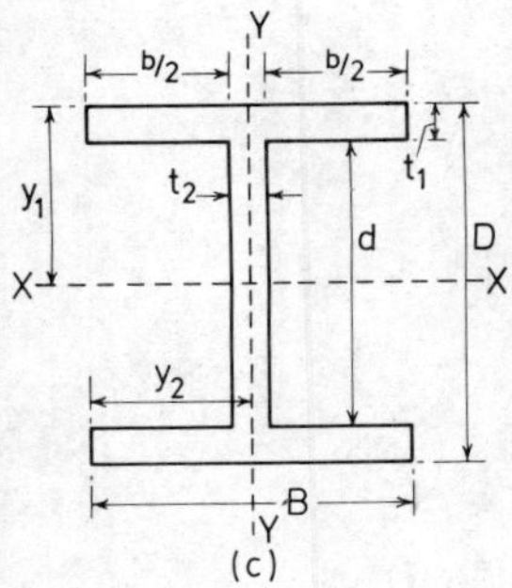

$$A = BD - bd$$

$$y_1 = \frac{D}{2} \quad y_2 = \frac{B}{2}$$

$$I_{XX} = \frac{BD^3 - bd^3}{12} \quad I_{YY} = \frac{2t_1B^3}{12} + \frac{dt_2^3}{12}$$

$$Z_{XX} = I_{XX} \div \frac{D}{2} \qquad Z_{YY} = I_{YY} \div \frac{B}{2}$$

$$r_{XX} = \sqrt{\frac{I_{XX}}{A}} \qquad r_{YY} = \sqrt{\frac{I_{YY}}{A}}$$

Table 1.4.2 (continued)

Rectangle with axis on base

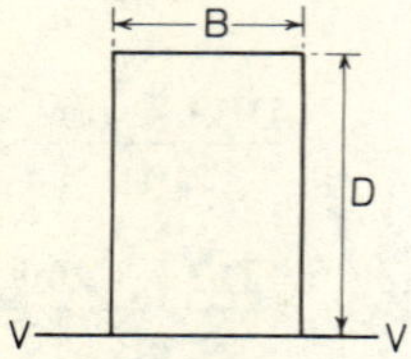

(d)

$$A = BD$$

$$I_{VV} = \frac{BD^3}{3}$$

Triangle with axis through centroid

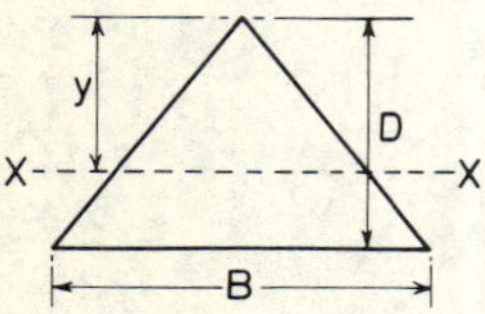

(e)

$$A = \frac{BD}{2}$$

$$y = \frac{2D}{3}$$

$$I_{XX} = \frac{BD^3}{36}$$

$$Z_{XX} = \frac{BD^2}{24}$$

$$r_{XX} = \frac{D}{\sqrt{18}} = 0{\cdot}236D$$

Table 1.4.2 (continued)

Triangle with axis on base or apex

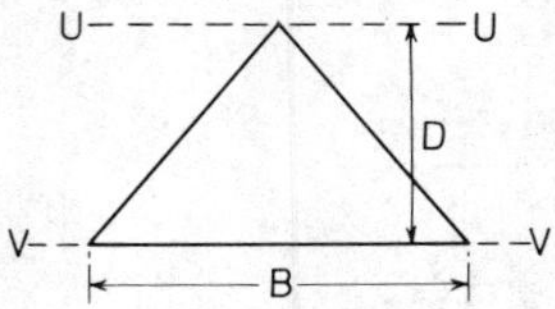

(f)

$$A = \frac{BD}{2}$$

$$I_{VV} = \frac{BD^3}{12} \quad I_{UU} = \frac{BD^3}{4}$$

Circle with axis through centroid

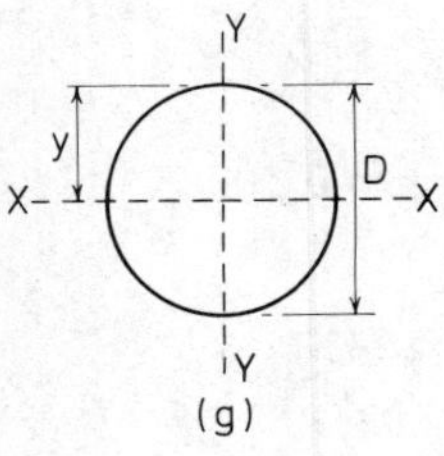

(g)

$$A = \frac{\pi D^2}{4}$$

$$y = \frac{D}{2}$$

$$I_{XX} = I_{YY} = \frac{\pi D^4}{64}$$

$$Z_{XX} = Z_{YY} = \frac{\pi D^3}{32}$$

$$r_{XX} = r_{YY} = \frac{D}{4} = 0{\cdot}25D$$

Table 1.4.2 (continued)

Hollow circle (i.e. ring)

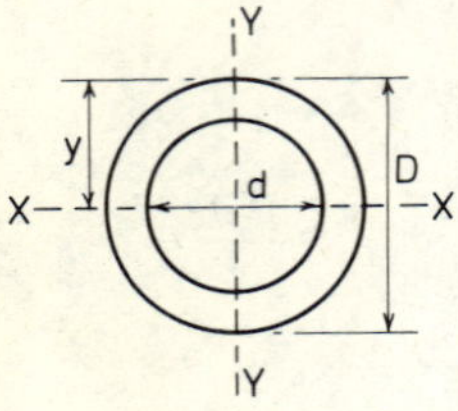

(h)

$$A = \frac{\pi(D^2 - d^2)}{4}$$

$$y = \frac{D}{2}$$

$$I_{XX} = I_{YY} = \frac{\pi(D^4 - d^4)}{64}$$

$$Z_{XX} = Z_{YY} = \frac{\pi(D^4 - d^4)}{32D}$$

$$r_{XX} = r_{YY} = \frac{\sqrt{(D^2 + d^2)}}{4}$$

$$= 0{\cdot}25\sqrt{(D^2 + d^2)}$$

Semicircle. Axis through centroid

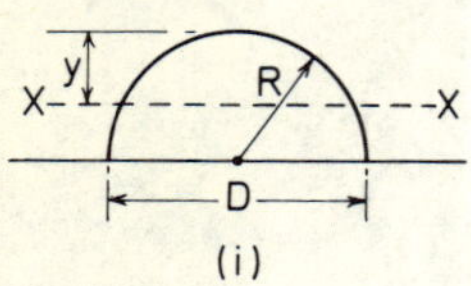

(i)

$$A = \frac{\pi D^2}{8}$$

$$y = R\left(1 - \frac{4}{3\pi}\right)$$

$$I_{XX} = R^4\left(\frac{\pi}{8} - \frac{8}{9\pi}\right) = 0{\cdot}1098R^4$$

$$r_{XX} = \sqrt{\frac{I_{XX}}{A}} = 0{\cdot}2643R$$

MOMENTS OF INERTIA OF UNSYMMETRICAL SECTIONS BUILT UP OF RECTANGLES

Usually the moments of inertia, etc., are required about axes which pass through the centroids (neutral axis). The first step, therefore, is to find the position of these axes and then the moments of inertia can be found by using the parallel axis theorem.

Parallel Axis Theorem

In Figure 1.4.1,

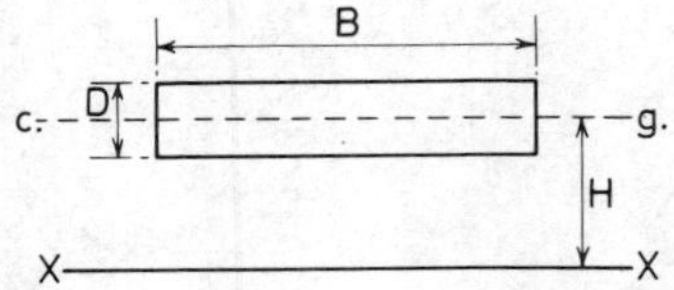

Figure 1.4.1

$$I_{XX} = I_{cg} + AH^2$$

Where I_{XX} = moment of inertia of figure about XX.

I_{cg} = moment of inertia about axis through centroid of section.

A = area of section.

H = distance between the parallel axes.

This formula is true for any shape of section.

For rectangles $I_{XX} = \frac{BD^3}{12} + BDH^2$ where B is the dimension parallel to axes c.g. and XX.

Example 1

In Figure 1.4.2,

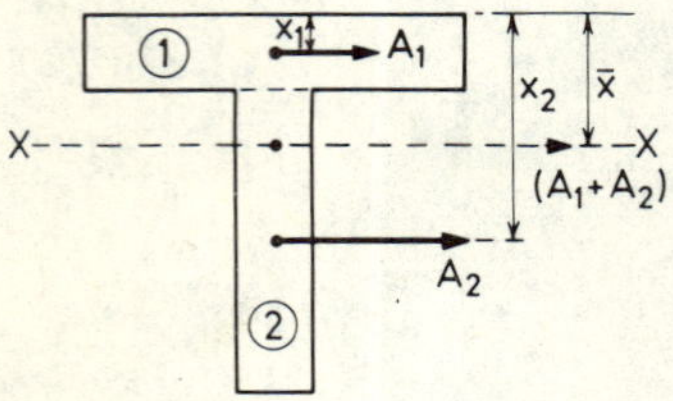

Figure 1.4.2

A_1 = area of rectangle ① acting through its c.g.

A_2 = area of rectangle ② acting through its c.g.

To find position of axis XX, i.e. to find value of $\bar{x}$, take moments about an extreme edge of figure.

$$\bar{x} = \frac{A_1x_1 + A_2x_2}{A_1 + A_2}$$

Having found position of axis XX use parallel axis theorem (see Figure 1.4.1).

Referring to Figure 1.4.3,

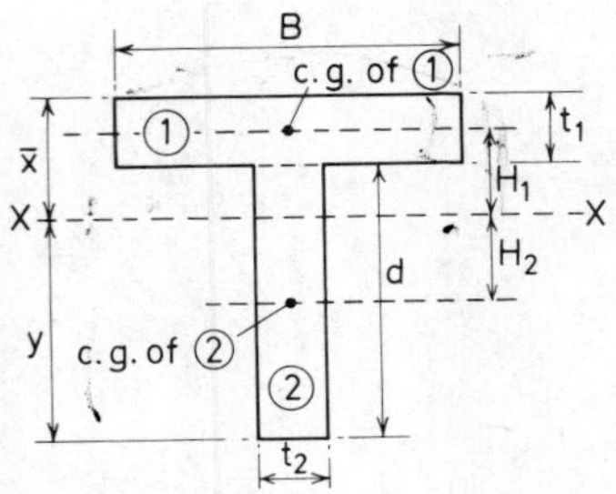

Figure 1.4.3

$$I_{XX} \text{ of } ① = \frac{Bt_1^3}{12} + Bt_1H_1^2$$

$$I_{XX} \text{ of } ② = \frac{t_2d^3}{12} + dt_2H_2^2$$

$$\text{Total } I_{XX} = I_{XX} \text{ of } ① + I_{XX} \text{ of } ②$$

Z_{XX} for use in beam design

$$= \frac{I_{XX}}{y}$$

Example 2

Referring to Figure 1.4.4,

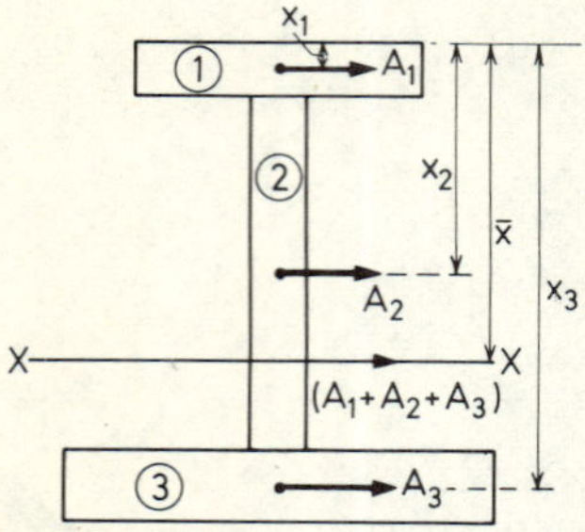

Figure 1.4.4

$$\bar{x} = \frac{A_1 x_1 + A_2 x_2 + A_3 x_3}{A_1 + A_2 + A_3}$$

From Figure 1.4.5,

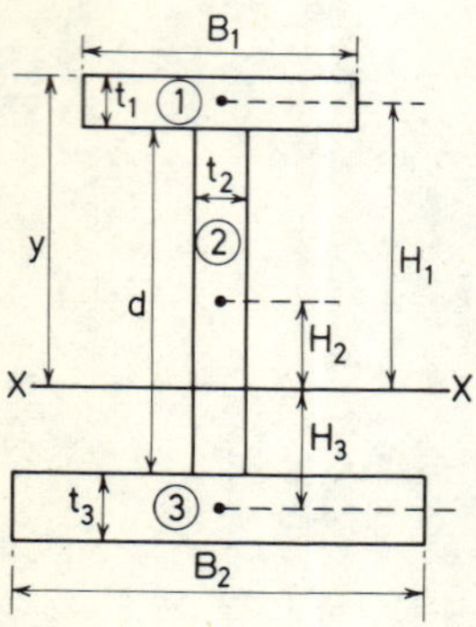

Figure 1.4.5

$$I_{XX} \text{ of } ① = \frac{B_1 t_1^3}{12} + B_1 t_1 H_1^2$$

$$I_{XX} \text{ of } ② = \frac{t_2 d^3}{12} + d t_2 H_2^2$$

$$I_{XX} \text{ of } ③ = \frac{B_2 t_3^3}{12} + B_2 t_3 H_3^2$$

$$\text{Total } I_{XX} = I_{XX} \text{ of } ① + I_{XX} \text{ of } ② + I_{XX} \text{ of } ③$$

Z_{XX} for use in beam design,

$$= \frac{I_{XX}}{y}$$

1.5 DEFINITIONS

Effective Length of Columns

See 'Slenderness Ratio'

Elasticity

An elastic material reverts to its original shape when a force which has caused small deformation is removed. Also, in an elastic material, stress is proportional to strain.

Factor of Safety

With regard to design of tension members, etc.

$$\text{Factor of safety} = \frac{\text{Ultimate or failing stress}}{\text{Permissible stress}}$$

(see also load factor). For materials such as mild steel which have a definite yield point, the factor of safety is sometimes taken as

$$\frac{\text{Yield stress}}{\text{Permissible stress}}$$

When making overturning calculations for walls the factor of safety against overturning is

$$\frac{\text{Resisting moment}}{\text{Overturning moment}}$$

For sliding of walls the factor of safety is

$$\frac{\text{Force resisting sliding}}{\text{Force causing sliding}}$$

Fixed-end Beams

With perfect fixity, the ends of the beam cannot move or turn. There are therefore bending moments at the fixed ends. The maximum bending moments are less than those in similar simply supported beams. When there is incomplete fixity, as in steel beams connected to stanchions by light angles, it is usual to calculate the bending moments as if the beams were simply supported.

Force

Force is any cause which produces or tends to produce motion or change of motion in the body upon which it acts. Force is measured in newtons.

Hooke's Law

In an elastic material, stress is proportional to strain.

Load Factor

$$\text{Load factor of a member} = \frac{\text{Load which would cause failure}}{\text{Design load}}$$

This of course gives a margin of safety, but load factor is not the same as factor of safety. Load factor design is based on actual conditions of stress at failure when stress is not proportional to strain, whereas the factor of safety, applied for example to beams, is usally based on the failing (or yield) stress in tension and compression and the beams designed assuming elasticity.

Middle Third Rule

This rule is usually employed in the design of masonry and mass concrete construction, e.g. foundations, piers, and retaining walls, and applies to rectangular cross-sections. In

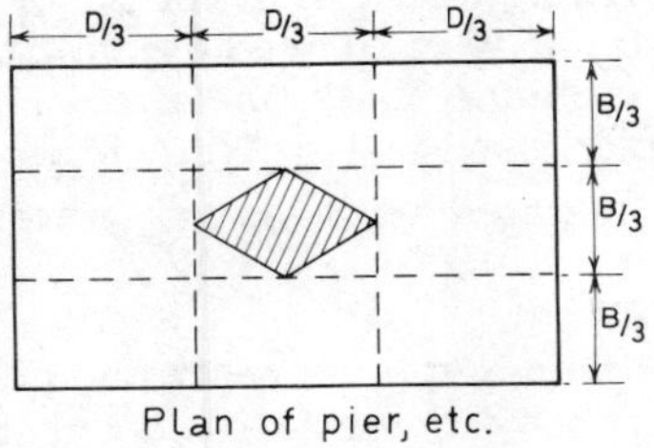

Figure 1.5.1

order that no tensile stresses may be developed in the material, the resultant thrust must not be outside the kern shown shaded in Figure 1.5.1.

Modulus of Elasticity or Young's Modulus

This refers to the lengthening or shortening of members subjected to force. Since in an elastic material stress is proportional to strain, stress divided by strain is a constant. This constant is called the modulus (i.e. measure) of elasticity and is usually denoted by E. It is expressed in kN/mm^2.

$$\frac{\text{Stress}}{\text{Strain}} = E \therefore \frac{W/A}{\delta/L} = E \text{ or } E = \frac{WL}{A\delta}$$

The modulus of elasticity gives an indication of the 'stiffness' of a material. The greater the value of E the more difficult it is to deform (i.e. lengthen or shorten or bend) the member, that is, a big stress is required to produce a small strain. The modulus of elasticity must not be confused with elastic modulus, which is printed in some structural steelwork tables instead of section modulus to distinguish between the elastic and plastic section moduli.

Moment of a Force

The moment of a force is the turning effect or leverage of that force about a given point or axis. It is measured by multiplying the force by the perpendicular distance of its line of action from the point or axis and is expressed in e.g., Nm, kNm, Nmm, etc.

Moment, Bending

When a member or portion of a member is so restrained or fixed than an applied moment cannot cause rotation about the point under consideration, bending is caused and the moment is then called a bending moment.

Moment of Inertia

This is a term first used in dynamics for rotating parts such as flywheels and armatures. A flywheel has inertia (i.e. a reluctance to having its state of rest or of motion changed) and a certain moment is required to cause it to rotate. This moment depends on the weight of the material and its arrangement with respect to the axis of rotation. The contribution of each particle of material to the total energy of the wheel depends on the weight of the particle and the square of its distance from the axis of rotation. Thus the moment of inertia is defined as the sum of all these products taking all particles into account, i.e., $I = \Sigma my^2$, where m = mass of the particles, y = distance from axis, and Σ means 'sum of'.

In deriving beam design formulae, the expression Σay^2 is arrived at where a is the area of a small element at a distance y from the axis of bending, and, because of its similarity with Σmy^2, it is called the moment of inertia, although it might be more accurate to call it the second moment of area.

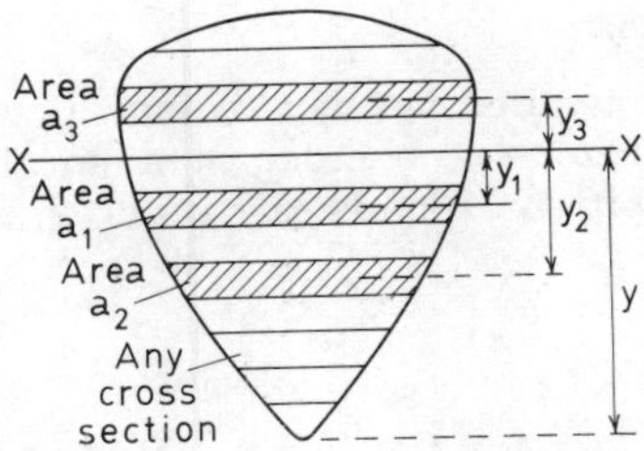

Figure 1.5.2

Referring to Figure 1.5.2,

$$I_{XX} = \Sigma ay^2$$
$$= a_1y_1^2 + a_2y_2^2 + a_3y_3^2 + a_4y_4^2 + \ldots, \text{etc.}$$

For geometrical figures, e.g. rectangles, circles, etc., general formulae for Σay^2 or I can be obtained by the use of the calculus. See Table 1.4.2.

Moment of Resistance

The bending moment in a beam must be resisted by an internal moment set up by the action of the beam fibres. The maximum bending moment (M) depends on the loads and reactions. The maximum safe moment of resistance (M_r) a beam can supply depends on the maximum permissible stress (f) for the material and the size and shape of the cross-section (i.e. the moment of inertia, I).

Normally, for design, the maximum moment of resistance (M_r) is made equal to the maximum bending moment (M) and the beam formula is $M = fI/y$, often expressed as $M = fZ$ (where $Z = I/y$). y is the distance from the neutral axis to the extreme fibre of the beam.

Pinned (i.e. Hinged-end) Members

A hinge keeps the ends of the member in position but allows freedom to turn. There is therefore no bending moment at a hinge. Depending on the disposition of loads, the reaction at a hinge can be at any angle.

Radius of Gyration

A term first used in dynamics, e.g. for flywheels. In comparing the effectiveness of such wheels, it is imagined that the whole weight of the wheel is concentrated into one particle so that the total moment of inertia is unaltered. The distance of this particle from the centre of the wheel is called the radius of gyration.

With respect to columns which are liable to buckle (bend) the radius of gyration can be thought of as the distance from the axis of bending at which the whole area of cross-section can be assumed to be concentrated so that the resistance to bending remains unaltered. The radius of gyration takes into account not only the size of the cross-section but its shape, that is the arrangement of the material with respect to the axis of buckling.

Usually, for the design of columns, the least radius of gyration is required (usually r_{YY}) where

$$\text{Least radius of gyration} = \sqrt{\frac{\text{Least moment of inertia}}{\text{Area of cross-section}}}$$

i.e. $$r_{YY} = \sqrt{\frac{I_{YY}}{A}}$$

Roller Bearing

When the end of a member is supported on rollers it is free to move longitudinally and also free to turn. The reaction at a roller bearing is always at right angles to the line of rollers.

Section Modulus

The moment of resistance (M_r) of a homogeneous beam depends on the nature of the material, and the size and shape of the cross-section. The nature of the material is provided for by the permissible stress (f) and the section modulus can be defined as the size-shape factor (Z).

$\therefore M_r = fZ$ where Z is the section modulus and $= I/y$.

Simply (or Freely) Supported Beams

A beam resting freely on walls (not built in) approximates closely to this condition. The ends of the beam are free to move longitudinally and are free to turn. There are, therefore, no bending moments at the supports.

Slenderness Ratio

The slenderness ratio of a column is usually expressed as

$$\frac{\text{Effective height of the column}}{\text{Least radius of gyration}} \text{ i.e. } \frac{l}{r}$$

Tables of permissible stresses for posts, struts, and columns are usually based on one type of end fixing of the column, i.e. 'Restrained at both ends in position but not in direction' (both ends hinged). In this case the effective height (l) of the column equals its actual height (L). For other conditions of end restraint allowance is made for the different load carrying capacities of columns by using effective heights or lengths as given in Table 1.5.1. This table is common to timber, reinforced concrete, and steel columns.

Table 1.5.1

Condition of end restraint	Effective length l
1. Restrained at both ends in position and direction (Both ends fixed)	0·7L
2. Restrained at both ends in position and one end in direction (One end fixed, one end hinged)	0·85L
3. Restrained at both ends in position but not in direction (both ends hinged)	L
4. Restrained at one end in position and direction and at the other end partially restrained in direction but not in position.	1·5L
5. Restrained at one end in position and direction but not restrained in either position or direction at the other end (One end fixed, one end free)	2·0L

Stress

Stress may be thought of as the internal 'distress' of a member as a result of the application of external loads. It is the resistance set up by the particles of the member in opposition to the breaking tendency of the loads. Stress is expressed as load per unit area (W/A), e.g. N/mm^2, kN/mm^2, MN/m^2.

In simple bending, stress = My/I, where M = bending moment, I = moment of inertia, and y = distance from the neutral axis to the fibre under consideration (usually the extreme fibre).

Strain

No member can be stressed without being deformed, i.e. strained. The most important strains are elongation (due to tension) and shortening (due to compression). Such strains are defined as alteration in length (δ) divided by original unloaded length (L) of the member.

$$\text{Strain} = \frac{\delta}{L}$$

Yield Stress

This applies particularly to mild steel. If a bar of steel is tested in tension, it behaves elastically until a certain stress (the yield stress) is reached. The bar then stretches a great deal and becomes more or less plastic. Increased load produces larger and larger extensions until finally the bar breaks.

Section 2

Structural Steelwork

Laterally Supported Beams
Laterally Unsupported Beams
Cased Beams
Columns, Stanchions and Struts
Angles in Tension
Rivets and Bolts
Welds
Stanchion Bases
Section Properties

2. Structural Steelwork

The code for the design of structural steelwork is BS 449; 'The Use of Structural Steel in Building.'

UNIVERSAL BEAMS AND COLUMNS

These have largely superseded the joists or British Standard Beams.

Dimensions and properties of universal beams and columns are given in Tables 2.9.1 to 2.9.3

2.1 LATERALLY SUPPORTED BEAMS

A laterally supported beam is a beam the compression flange of which is restrained from buckling by virtue of the lateral support given by floors or other beams.

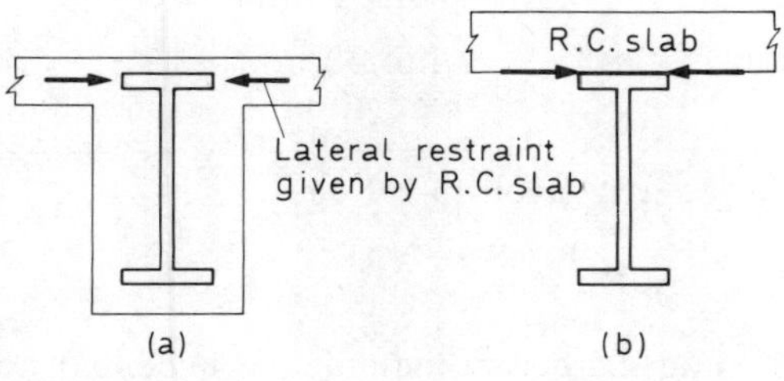

Figure 2.1.1

There is no need to consider buckling for the beam of Figure 2.1.1a. In Figure 2.1.1b the friction supplied by the slab is usually sufficient to eliminate the need for sideways buckling calculations.

Simple Design

In the 'simple' method of design referred to in BS 449, beams connected to stanchions or other beams by simple angle cleats and brackets are assumed to be simply supported for the calculation of bending moments.

Design of Laterally Supported Beams

p_{bc} = permissible stress in compression due to bending

Table 2.1.1 Allowable stress p_{bc} in bending

Section	Grade	Thickness of material	p_{bc} (N/mm^2)
Rolled I-beams and channels	43	all	165
Plates, angles, tees	43	up to 40 mm over 40 mm	165 150
Universal beams and columns	43	up to 40 mm over 40 mm	165 150

Values for other grades are given in BS 449.

Design

1. Estimate the load on the beam, including if necessary an allowance for the weight of the beam and its concrete casing (if any).
2. Calculate the maximum bending moment M in N mm.
3. $Z = M/p_{bc}$. Z is in mm^3, and as the available properties are in cm^3, the calculated value of Z must be then divided by 1000. The value of Z in the tables which is slightly

greater than that calculated then gives the required section. (For many students this is the final answer, but checks for shear, buckling and deflection may be needed.)

4. Shear

Average shear stress $= \dfrac{\text{Shear force}}{\text{Dt}}$ (see Figure 2.1.2)

This calculated shear stress should not exceed 115 N/mm^2 for grade 43 steel up to 40 mm thick or 105 N/mm^2 for grade 43 steel greater than 40 mm thick.

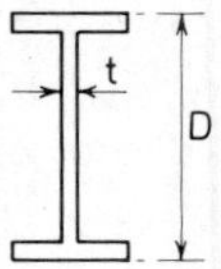

Figure 2.1.2

5. Buckling

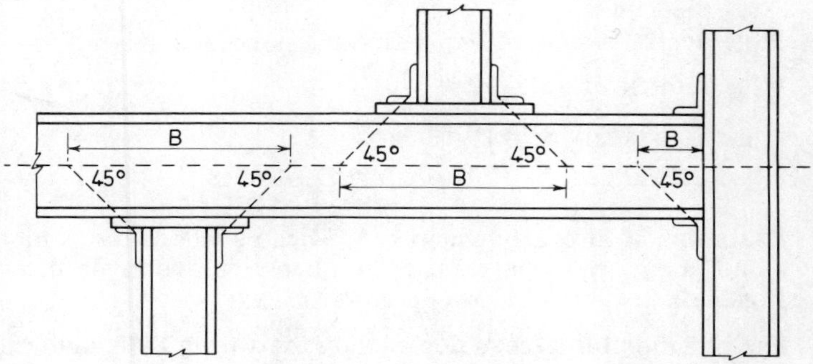

Figure 2.1.3

In Figure 2.1.3, t = web thickness and d = clear depth of web between root fillets. The 45° lines are taken from the outside edge of the stiff portion of the bearing.

Bearing stiffeners are required at point of concentrated loads and at supports when the concentrated load or reaction exceeds $p_c tB$, where p_c = axial stress for struts for a slenderness ratio of $(d/t).\sqrt{3}$.

6. Deflection
   Maximum deflection should not exceed 1/360, due to loads other than the weight of the structural floors or roof, steelwork, and casing.

If the beam carries a U.D. load, this requirement is satisfied if span/depth ratios for sections of grade 43 steel do not exceed the values in Table 2.1.2.

Table 2.1.2

W_d/W_t	1·0	0·9	0·8	0·7	0·6	0·5
Span/depth ratio	17	15·4	13·6	11·8	10·2	8·5

W_d = load considered for deflection purposes

W_t = total load

2.2 LATERALLY UNSUPPORTED BEAMS

Examples of laterally unsupported beams where the compression flange is not restrained from buckling are given in Figure 2.2.1.

The permissible stress depends on the ratios D/T and l/r_Y, where l = effective length of the compression flange and r_Y = radius of gyration about axis YY.

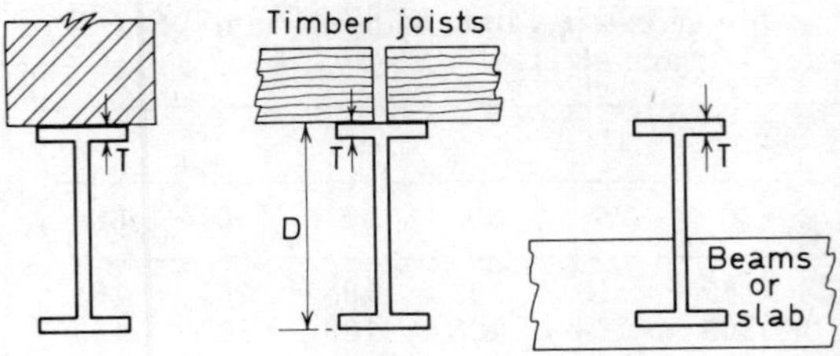

Figure 2.2.1

The effective lengths l for two simple types of connections are given in Figure 2.2.2.

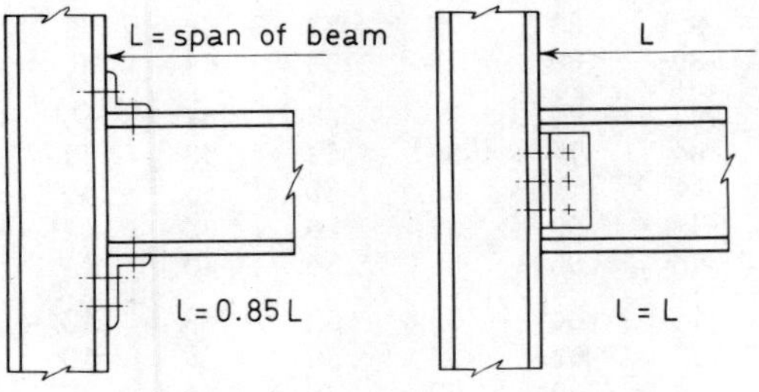

Figure 2.2.2

When l/r_Y does not exceed 85 for grade 43 steel the stresses given in Table 2.1.1 for laterally supported beams may be used.

Table 2.2.1 gives permissible stresses for grade 43 steel only. See BS 449 for a similar table for other grades.

Table 2.2.1 Allowable stress p_{bc} in bending (N/mm^2) for beams of grade 43 steel

l/r_Y	D/T							
	10	15	20	25	30	35	40	50
90	165	165	165	165	165	165	165	165
95	165	165	165	163	163	163	163	163
100	165	165	165	157	157	157	157	157
105	165	165	160	152	152	152	152	152
110	165	165	156	147	147	147	147	147
115	165	165	152	141	141	141	141	141
120	165	162	148	136	136	136	136	136
130	165	155	139	126	126	126	126	126
140	165	149	130	115	115	115	115	115
150	165	143	122	104	104	104	104	104
160	163	136	113	95	94	94	94	94
170	159	130	104	91	85	82	82	82
180	155	124	96	87	80	76	72	71
190	151	118	93	83	77	72	68	62
200	147	111	89	80	73	68	64	59
210	143	105	87	77	70	65	61	55
220	139	99	84	74	67	62	58	52
230	134	95	81	71	64	59	55	49
240	130	92	78	69	61	56	52	47
250	126	90	76	66	59	54	50	44
260	122	88	74	64	57	52	48	42
270	118	86	72	62	55	50	46	40
280	114	84	70	60	53	48	44	39
290	110	82	68	58	51	46	42	37
300	106	80	66	56	49	44	41	36

Intermediate values may be obtained by linear interpolation.

Note. For materials over 40 mm thick the stress shall not exceed 150 N/mm^2.

Design of Laterally Unsupported Beams (Mild Steel)

1. Choose a trial beam from the tables and read off the values of Z_{xx} and D/T and r_y. Calculate l/r_y. Convert dimensions to millimetres.

2. From Table 2.2.1 read off the permissible stress p_{bc} and calculate the moment of resistance (M_r) of the beam from $M_r = p_{bc}Z_{xx}$.

3. Calculate the maximum bending moment (M) due to the loads. If M_r is equal to, or not much greater than M, the beam is suitable for bending. If not, choose another section and repeat the calculations.

4. If necessary, check for shear, buckling and deflection.

2.3 CASED BEAMS (GRADE 43 STEEL)

See Figure 2.3.1. r_y may be taken as 0·2 (B + 100 mm).

D/T is the same as for the uncased section. The permissible stress obtained from Table 2.2.1 must not exceed $1\frac{1}{2}$ times that permitted for the uncased section, nor must it exceed the maximum permissible stress given in Table 2.1.1 for laterally supported beams.

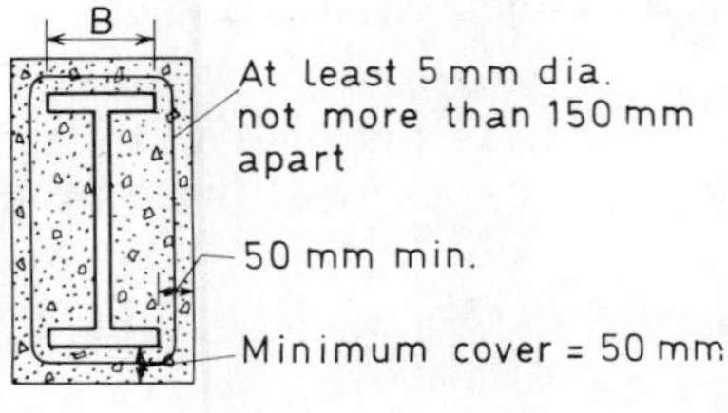

Figure 2.3.1

2.4 COLUMNS, STANCHIONS, AND STRUTS

Table 2.4.1 gives the permissible stresses for grade 43 steel where p_c = permissible axial stress in compression on the gross cross section of the steel strut, (N/mm^2) = effective height of stanchion (mm), and r = appropriate radius of gyration (usually the least radius of gyration which for I sections is r_y) in mm.

Stresses for other grades are given in BS 449.

Maximum Slenderness Ratio of Struts

For any member carrying dead loads with or without imposed loads	180
For any member carrying loads resulting from wind forces only	250
For any member normally acting as a tie in a roof truss but subject to reversal of stress resulting from the action of wind	350

Axial Loads on Struts, i.e. Columns, Stanchions, Etc.

To design a universal uncased column to carry a known axial load,

1. Choose a trial column from tables and read off its area A and least radius of gyration r_y. Convert these values to mm^2 and mm.

2 Calculate the slenderness ratio l/r_y where l is the effective length of the column in millimetres.

3. From Table 2.4.1 read off the appropriate safe axial stress p_c.

Table 2.4.1 Allowable stress p_c on gross section for axial compression

	p_c (N/mm²) for grade 43 steel									
l/r	0	1	2	3	4	5	6	7	8	9
0	155	155	154	154	153	153	153	152	152	151
10	151	151	150	150	149	149	148	148	148	147
20	147	146	146	146	145	145	144	144	144	143
30	143	142	142	142	141	141	141	140	140	139
40	139	138	138	137	137	136	136	136	135	134
50	133	133	132	131	130	130	129	128	127	126
60	126	125	124	123	122	121	120	119	118	117
70	115	114	113	112	111	110	108	107	106	105
80	104	102	101	100	99	97	96	95	94	92
90	91	90	89	87	86	85	84	83	81	80
100	79	78	77	76	75	74	73	72	71	70
110	69	68	67	66	65	64	63	62	61	61
120	60	59	58	57	56	56	55	54	53	53
130	52	51	51	50	49	49	48	48	47	46
140	46	45	45	44	43	43	42	42	41	41
150	40	40	39	39	38	38	38	37	37	36
160	36	35	35	35	34	34	33	33	33	32
170	32	32	31	31	31	30	30	30	29	29
180	29	28	28	28	28	27	27	27	26	26
190	26	26	25	25	25	25	24	24	24	24
200	24	23	23	23	23	22	22	22	22	22
210	21	21	21	21	21	20	20	20	20	20
220	20	19	19	19	19	19	19	18	18	18
230	18	18	18	18	17	17	17	17	17	17
240	17	16	16	16	16	16	16	16	16	15
250	15									
300	11									
350	8									

Intermediate values may be obtained by linear interpolation.

Note. For material over 40 mm thick, other than rolled I-beams or channels, and for Universal columns of thicknesses exceeding 40 mm, the limiting stress is 140 N/mm².

4. The safe axial load = p_cA. If this load is less or appreciably greater than the design load, choose another column and repeat the calculations.

Note. The effective height can be obtained from consideration of Table 1.5.1. A rough practical guide is given in Figure 2.4.1.

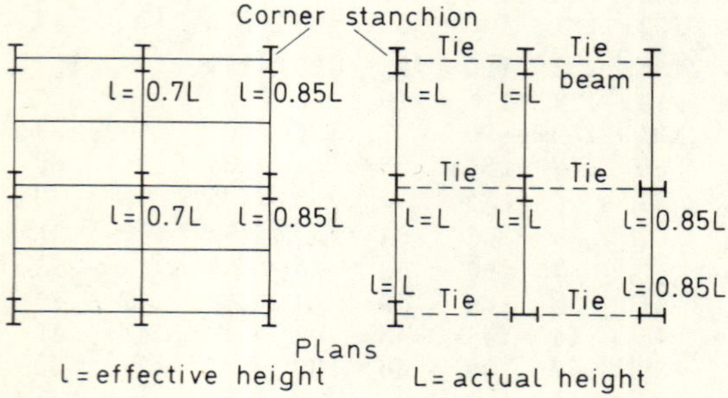

Figure 2.4.1

To Design a Solid Round Steel Column for an Axial Load

Proceed as above for uncased universal columns except that $r_y = D/4$ and $A = \pi D^2/4$ where D = diameter of column.

To Design an Axially Loaded Cased Stanchion of I-section (Mild Steel).

See Figure 2.4.2. The following notes apply to mild steel columns of overall dimensions not exceeding 500 × 1000 mm. The radius of gyration r_y can be taken as 0·2 (B + 100 mm). (This applies even if the cover exceeds 50 mm).

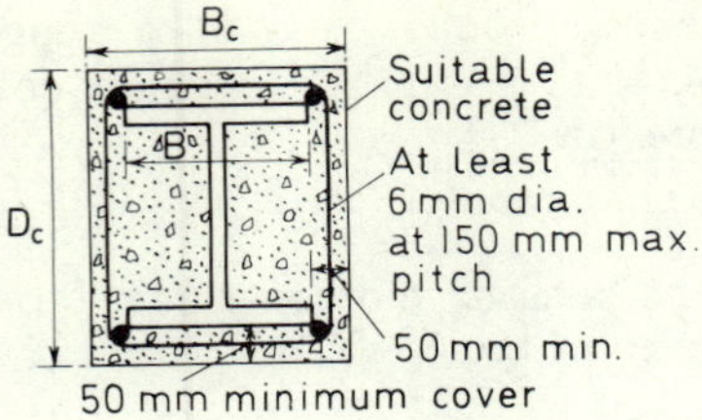

$\gamma_r = 0.2(B + 100\,\text{mm})$.

Figure 2.4.2

Slenderness ratio $= l/r_y$.

$$\text{Safe axial load} = p_c A + \frac{p_c}{0{\cdot}19 p_{bc}} B_c D_c. \qquad (1)$$

A = cross-sectional area of steel stanchion

$B_c D_c$ = overall cross-sectional area. Cover in excess of 75 mm to be ignored.

The safe axial load obtained from (1) must not exceed twice that which would be permitted on the uncased section. For cased columns of high yield steel see BS 449 and Amendments.

Eccentric Loading on Steel Stanchions

The amount of eccentricity to be assumed for some typical connections is given in Figure 2.4.3.

For continuous stanchions (such as in multistorey buildings) the bending moments due to eccentricities of loading at any one floor may be taken as being ineffective at the floors above and below that floor. The bending moments at any floor level may be equally divided between the stanchion lengths above and below the floor providing the ratio of their stiffnesses,

K_1/K_2 (or K_2/K_1) does not exceed 1·5.

$$K = \frac{I}{L} = \frac{\text{Moment of inertia}}{\text{Actual length of stanchion}}$$

Where the ratio exceeds 1·5 the bending moments are to be divided in proportion to the ratio of the K-values.

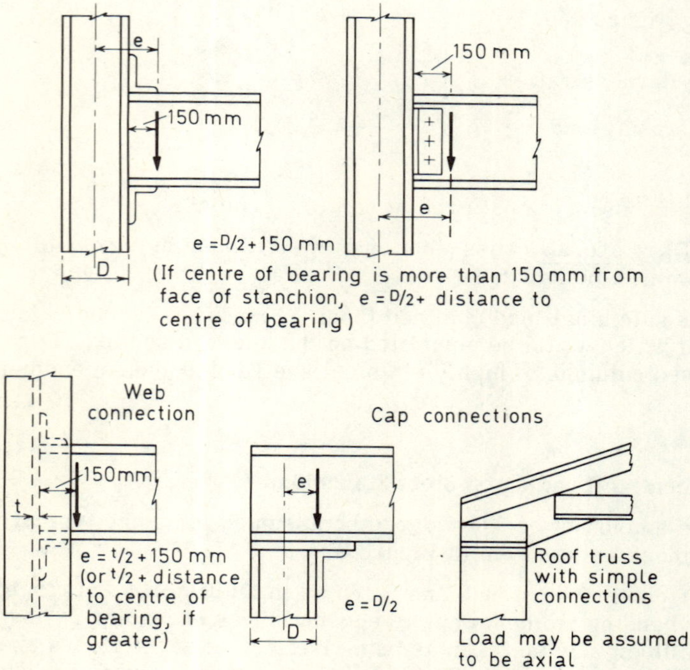

Figure 2.4.3

To Design an Uncased Stanchion

See Figure 2.4.4 where stanchion length marked ① is to be designed.

1. Choose a trial section.

2. Calculate its slenderness ratio l/r_y and read off the permissible stress p_c from Table 2.4.1.

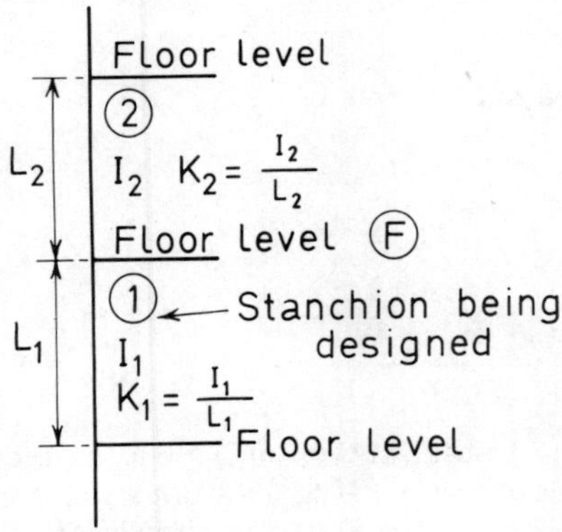

Figure 2.4.4

3. Add together all the loads, including the eccentric loads at the floor level (F) and the loads from the floors above, and calculate the direct axial compressive stress f_c where

$$f_c = \frac{\text{Total load}}{\text{Area of cross-section of the stanchion}}$$

4. Calculate the maximum compressive stress f_{bc} due to the eccentric loads at the floor level (F), making allowance for the division of bending moments between the stanchion

being designed (Stanchion ①) and the next length of stanchion above (Stanchion ②). Two examples are given below.

5. The stanchion is safe if

$$\frac{\text{Actual direct stress}}{\text{Permissible direct stress}} + \frac{\text{Actual bending stress}}{\text{Permissible bending stress}}$$

does not exceed unity,

i.e. if $\dfrac{f_c}{p_c} + \dfrac{f_{bc}}{p_{bc}}$ does not exceed 1.

Example 1

Referring to Figure 2.4.5,

$$\text{Direct stress} = f_c = \frac{W + W_x + W_y}{A}$$

(A = area of cross-section of steel stanchion.) Bending moment about axis XX = $M_x = 0{\cdot}5\ W_x e_x$ if K_1/K_2 does not exceed 1·5 or $(K_1/K_1 + K_2)W_x e_x$ if the K ratio exceeds 1·5.

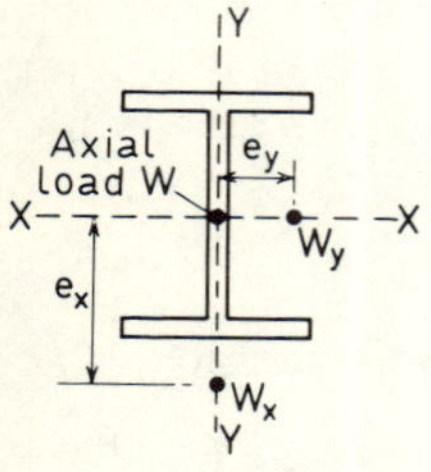

Figure 2.4.5

Bending moment about axis YY = M_y

$$= 0{\cdot}5\, W_y e_y \text{ if the K ratio does not exceed } 1{\cdot}5$$

$$= \frac{K_1}{K_1 + K_2} W_y e_y \text{ if the K ratio exceeds } 1{\cdot}5$$

Maximum compressive stress due to bending

$$= f_{bc} = \frac{M_x}{Z_{xx}} + \frac{M_y}{Z_{yy}}$$

The stanchion is safe if $\frac{f_c}{p_c} + \frac{f_{bc}}{p_{bc}}$ does not exceed unity.

Example 2

It is assumed in this example, Figure 2.4.6, that the stanchion is the top length supporting the roof therefore all the bending moments are taken by this length.

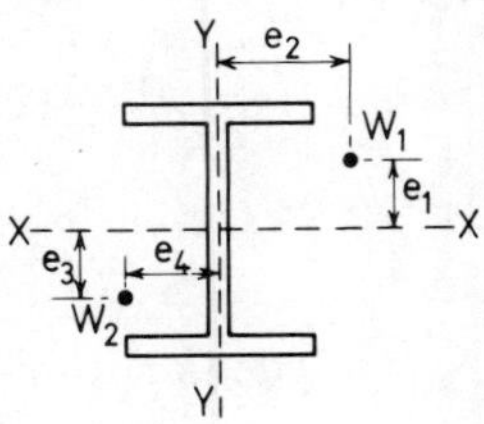

Figure 2.4.6

1. Choose a trial section.
2. Calculate l/r_y and obtain p_c.

3. $f_c = \dfrac{W_1 + W_2}{A}$

4. $f_{bc} = \dfrac{M_x}{Z_{xx}} + \dfrac{M_y}{Z_{yy}}$

where $M_x = W_1e_1 - W_2e_3$ (assuming W_1e_1 is greater than W_2e_3) and $M_y = W_1e_2 - W_2e_4$ (assuming W_1e_2 is greater than W_2e_4).

5. The stanchion is safe if $f_c/pc + f_{bc}/p_{bc}$ does not exceed unity.

Cased Stanchions

Bending moments due to eccentric loading are deemed to be taken by the steel core alone.

1. Choose a trial section.
2. Calculate its slenderness ratio l/r_y where $r_y = 0{\cdot}2$ (B + 100) mm and read off p_c from Table 2.4.1.
3. Add together all the loads including the eccentric loads, giving total load W_t.
   Calculate the allowable axial load W_a from

$$W_a = p_cA + \frac{p_c}{0{\cdot}19p_{bc}} B_cD_c$$

4. Calculate f_{bc} as for uncased stanchions.
5. The stanchion is safe if $W_t/W_a + f_{bc}/p_{bc}$ does not exceed unity.

Angles as Struts

Single-angle discontinuous struts connected to gussets or to a section by not less than two rivets or bolts at each end or by their equivalent in welding (Figure 2.4.7).

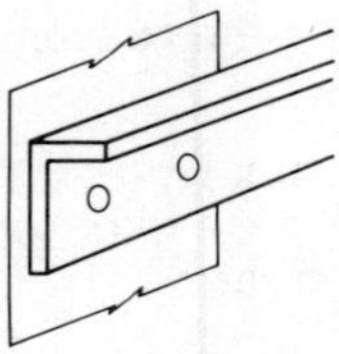

Figure 2.4.7

Design as an axially loaded member.

Take l as 0·85 times the length of the strut centre to centre of interesections at each end and r (from tables) as the minimum radius of gyration.

Calculate the slenderness ratio l/r and look up permissible stress p_c in Table 2.4.1.

Safe load = $p_c A$ where A = area of cross-section of angle.

Double-angle discontinuous struts as shown in Figure 2.4.8 if there are not less than two rivets or bolts in line at each end.

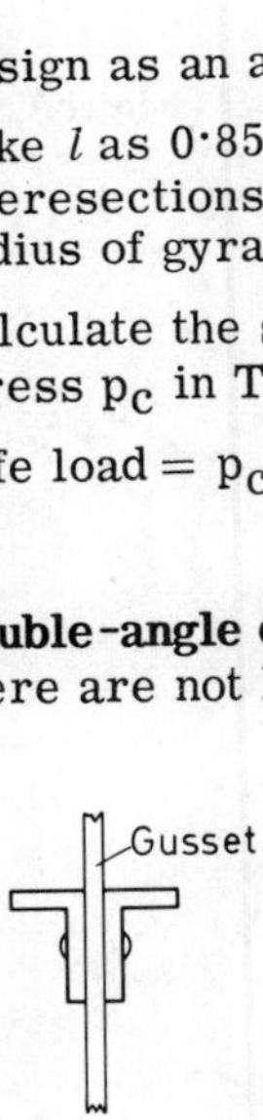

Figure 2.4.8

l = 0·7 to 0·85 times the distance between intersections.

r = minimum radius of gyration. between these 2 angles

p_c = permissible stress from Tables 2.4.1 for the appropriate value of l/r.

Safe load = $p_c A$, where A = area of cross-section (two angles). Solid washers or packings must be used at not less than two places spaced equidistant in the length of the strut.

2.5 ANGLES IN TENSION

Single Angle

Effective area A for calculation of tensile strength (Figure 2.5.1) is given by

$$A = a_1 + a_2 \left(\frac{3a_1}{3a_1 + a_2} \right)$$

where a_1 = net sectional area of the connected leg = $t(d_1 - t/2)$ minus area of rivet or bolt hold, and

a_2 = sectional area of the unconnected leg = $t(d_2 - t/2)$.

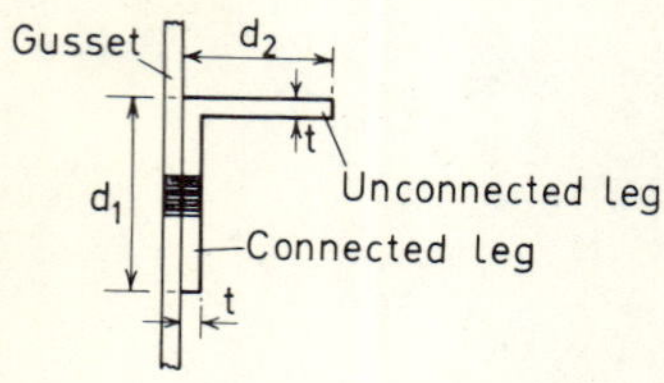

Figure 2.5.1

Safe tensile load = $p_t A$, where p_t = 155 N/mm^2 for grade 43 steel sections except those sections over 40 mm, where p_t = 140 N/mm^2.

Double Angle See Figure 2. 5. 2

Effective area A = gross area of the two angles minus area of rivet or bolt hole.

Safe tensile load = p_tA.

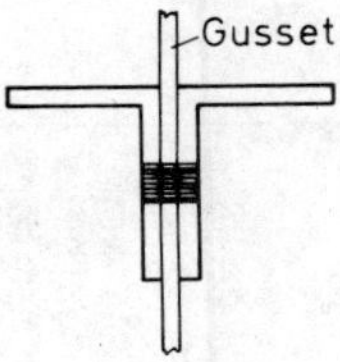

Figure 2. 5. 2

2. 6 RIVETS AND BOLTS

For rivets and bolts in double shear the area to be assumed shall be twice the area for single shear. Where the rivets or bolts are in single shear, the permissible bearing stress shall be reduced by 20 per cent

Single Shear

S.S.V. of one rivet = f_sA where A = gross area = $\pi d^2/4$ where d = gross diameter.

B.V. of one rivet = $0{\cdot}8\ f_bdt$.

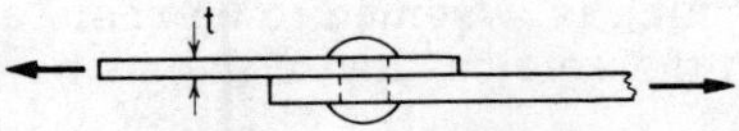

Figure 2. 6. 1

Table 2.6.1 Allowable stresses in rivets and bolts in N/mm^2 on gross areas.

Description		Bolts of strength grade designation 4·6	Rivets of material having a yield stress in N/mm^2 of 250	350	430
In shear:	power-driven shop rivets		110	155	190
	power-driven field rivets		100	140	170
	hand-driven rivets		90	125	155
	close tolerance turned bolts	95			
	black bolts	80			
In bearing:	power-driven shop rivets		315	440	540
	power-driven field rivets		290	405	500
	hand-driven rivets		265	370	455
	close tolerance turned bolts	300			
	black bolts	200			

The gross area of a rivet is equal to the rivet hole.

Double Shear

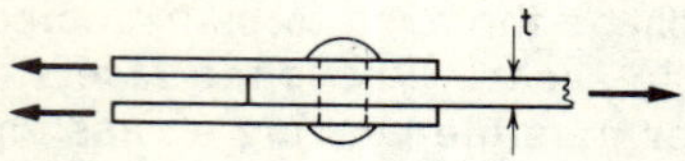

Figure 2.6.2

D.S.V. of one rivet $= 2 f_s A$.

B.V. of one rivet $= f_b dt$.

Beam-to-stanchion Connections

W = load from end of beam which is assumed to be resisted entirely by the shop rivets in the vertical leg of the seating bracket.

$$\text{Number of rivets required} = \frac{W}{\text{Safe load for one rivet}}$$

Single shear value or bearing value } whatever is smaller

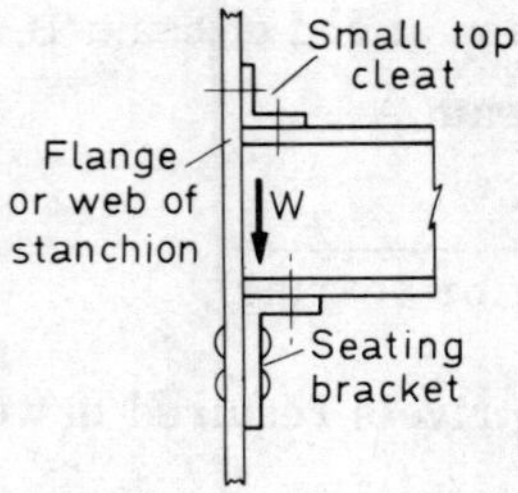

Figure 2.6.3

Usually the safe load for one rivet in these connections is the single shear value. Use however the bearing value in the angle or flange or web if it is less than the shear value.

Beam-to-beam Connections

Example 1

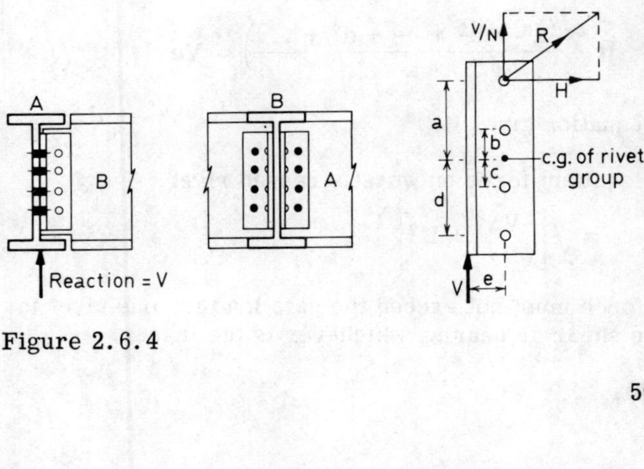

Figure 2.6.4

Referring to Figure 2. 6. 4, V = reaction at end of beam B.

Number of bolts required in web of beam A

$$= \frac{V}{\text{Safe load for one bolt in shear (or bearing)}}$$

Ignoring eccentricity, number of shop rivets required in web of beam B

$$= \frac{V}{\text{Safe load for one rivet}}$$

where the safe load is the lesser of the double shear value of one rivet or bearing in the web of beam B.

If eccentricity is taken into account, assume a number of rivets and let a be the distance of the rivet furthest away from the centre of gravity of the rivet group. b, c, d, etc. are the distances from the c.g. of the other rivets. Vertical load on each rivet = V/N where N = number of rivets.

Let H be the horizontal force on the worst stressed rivet due to the moment Ve.

Then $$H \times \left(\frac{a^2 + b^2 + c^2 + d^2 + \ldots}{a}\right) = Ve$$

This equation gives H.

Resultant force on worst stressed rivet

$$= \sqrt{\left[\left(\frac{V}{N}\right)^2 + H^2\right]}$$

This force must not exceed the safe load for one rivet in double shear or bearing whichever is the lesser.

Example 2

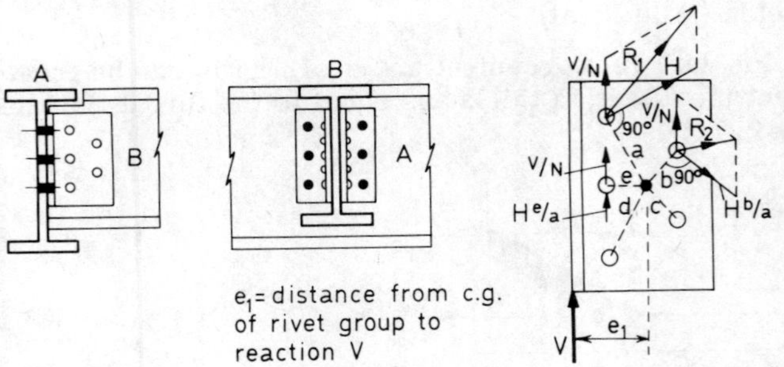

Figure 2.6.5

Number of bolts in web of beam A as for Example 1.

Assume a grouping of rivets for web B, obtain the position of the c.g. of the group, and the distances a, b, c, etc. from the c.g. Vertical force on each rivet = V/N where N is the number of rivets.

The force on each rivet due to the moment Ve_1 will be at right angles to the line joining the rivet to the c.g. of the group. Let H be the force due to the moment on the rivet furthest away (distance a) from the c.g. (The forces on the other rivets will be H^b/a, H^c/a, etc.)

$$H \times \left(\frac{a^2 + b^2 + c^2 + d^2 + e^2}{a}\right) = Ve_1.$$

This equation gives H.

The resultant force (R_1, R_2 etc.) on each rivet can be obtained by the parallelogram of forces as shown in Figure 2.6.4. The resultant force on the worst-stressed rivet must not exceed the safe load for one rivet in double shear or bearing, whichever is the lesser.

2.7 WELDS

Butt Welds (Mild Steel)

Butt welds with reinforcement and sealing run can be treated as parent metal with a thickness equal to the throat thickness (Figure 2.7.1).

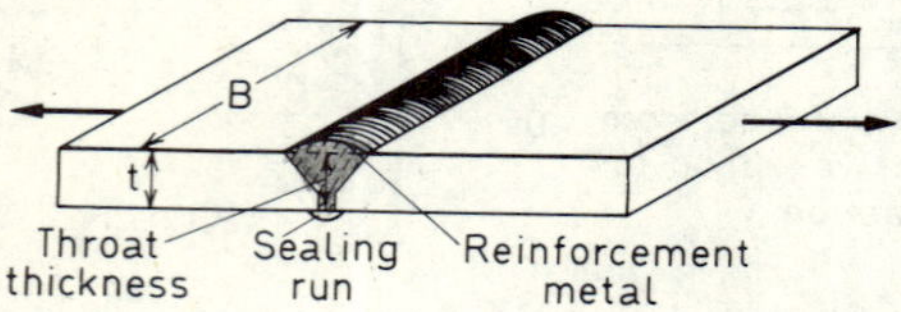

Figure 2.7.1

The permissible stresses are then equal to those for the parent metal. If the plate shown in Figure 2.7.1 is in tension, the safe load (mild steel) is $p_tA = 155Bt$ newtons.

Table 2.7.1 Strength in kilonewtons per ten millimetre (10 mm) run of right-angle weld at 115 N/mm^2

Size (mm)	Strength (kN)	Size (mm)	Strength (kN)
4	3·22	14	11·37
6	4·83	16	12·88
8	6·44	18	14·49
10	8·05	20	16·10
12	9·66	25	20·13

Fillet Welds

Permissible stress for grade 43 steel = 115 N/mm^2. For right-angle welds as shown the throat thickness = 0·7 × size.

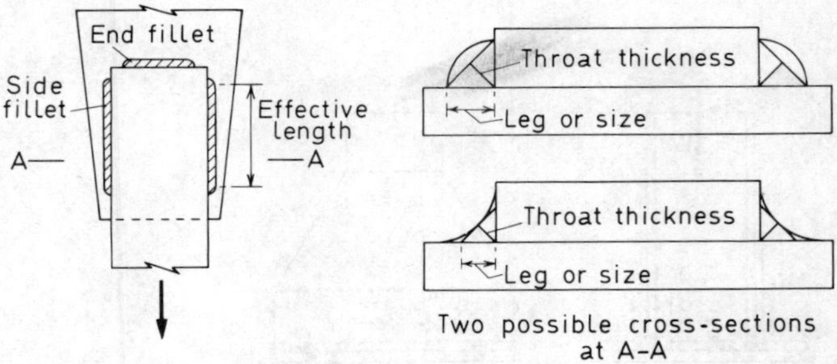

Figure 2.7.2

Permissible load for ten millimetre run of right-angle weld = 10 × 0·7 × size × 115 newtons = 805 × size (N).

Permissible load for a whole run of weld = 80·5 × size (mm) × effective length (mm) newtons.

(Effective length = overall length of weld minus twice weld size.)

2.8 STANCHION BASES

Gusset bases

For stanchions with axial loads where all parts are fabricated flush for bearing (approximate method).

1. Number of rivets in flanges of stanchion

$$= \frac{0{\cdot}6\,W}{\text{S.S.V. of one rivet}}$$

i.e. the rivets are designed for 60% of the load W on the stanchion.

2. Minimum area of base plate = W divided by the permissible pressure on the concrete or reinforced concrete foundation (Table 2.8.1).

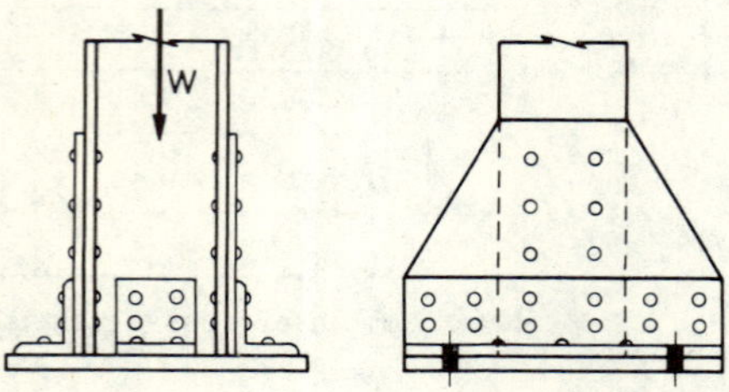

Figure 2.8.1

Table 2.8.1

Nominal mix	Permissible pressure (kN/m^2)
1 : 1 : 2	7300
1 : 1½ : 3	6100
1 : 2 : 4	4900

3. Thickness of base plate. In Figure 2.8.2,

t = combined thickness of base plate and angle

L_1L_2 = area of base = A

$$\text{Upthrust on shaded area} = \frac{W}{L_1L_2} \times L_1x = \frac{Wx}{L_2}$$

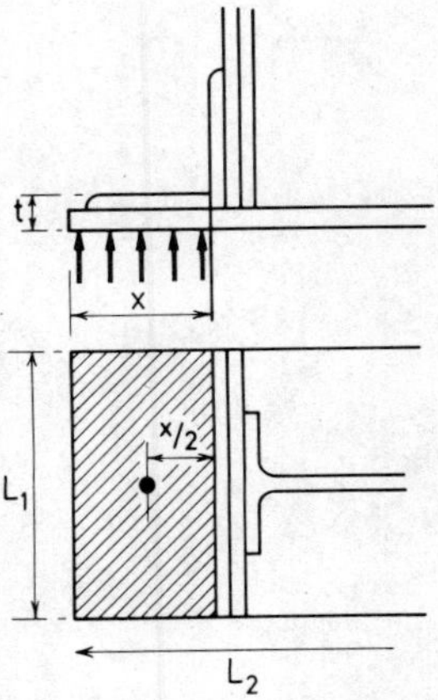

Figure 2.8.2.

$$\text{Bending moment } M = \frac{Wx^2}{2L_2}$$

$$t = \sqrt{\left(\frac{6Wx^2}{2L_1L_2f}\right)} = \sqrt{\left(\frac{Wx^2}{4L_1L_2}\right)} = \frac{x}{2}\sqrt{\frac{W}{A}}$$

when $f = 185$ N/mm^2.

Slab Bases

In Figure 2.8.3,

$$t = \sqrt{\left[\frac{3W}{p_{bct}}\left(A^2 - \frac{B^2}{4}\right)\right]}$$

t = slab thickness (mm)

A = greater projection (mm)

B = lesser projection (mm)

$$w = \frac{W}{L_1 L_2}\ \text{N/mm}^2$$

p_{bct} = permissible bending stress = 185 N/mm^2.

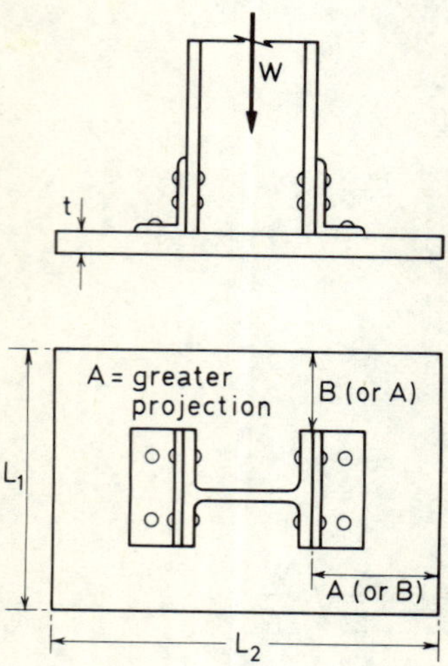

Figure 2.8.3

Slab Bases for Solid Round Columns

In Figure 2.8.4,

$$t = \sqrt{\left(\frac{9W}{16p_{bct}}\frac{D}{D-d}\right)}$$

t = slab thickness (mm)

W = total axial load (N)

D = length of side of square base (mm)

d = diameter of column = diameter of reduced end, if any (mm)

p_{bct} = 185 N/mm^2

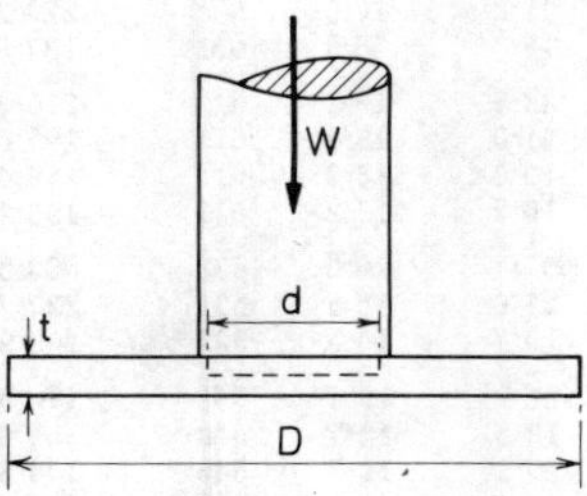

Figure 2.8.4

2.9 SECTION PROPERTIES

Tables 2.9.1 to 2.9.8 give the properties of some steel sections, mainly in centimetre units. International sizes in metric units are expected to be agreed, but in the meantime steel tables are available giving safe loads etc. for imperial sections in metric units.

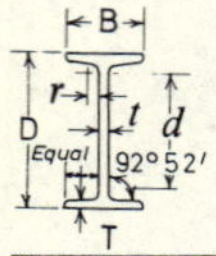

Table 2.9.1

Universal Beams

Dimensions and Properties

Serial Size	Weight per foot	Weight per metre	Depth of Section D	Width of Section B	Thickness Web r	Thickness Flange T	Root Radius r	Depth between Fillets d	Area of Section
m	lb	kg	mm	mm	mm	mm	mm	mm	cm²
36 × 16½	260	387	920	420	21·5	36·6	24·1	791	493·9
	230	343	911	418	19·4	32·0	24·1	791	436·9
36 × 12	194	289	927	308	19·6	32·0	19·1	819	368·5
	170	253	918	305	17·3	27·9	19·1	819	322·5
	150	223	910	304	15·9	23·9	19·1	819	284·9
	135	201	903	303	15·2	20·2	19·1	819	256·1
33 × 11½	152	226	851	294	16·1	26·8	17·8	757	288·4
	130	194	841	292	14·7	21·7	17·8	757	246·9
	118	176	835	291·6	14·0	18·8	17·8	757	223·8
30 × 10½	132	196	770	268	15·6	25·4	16·5	681	250·5
	116	173	762	267	14·3	21·6	16·5	681	220·2
	99	147	754	265	12·9	17·5	16·5	681	187·8
27 × 10	114	170	693	256	14·5	23·7	15·2	611	216·3
	102	152	688	254·5	13·2	21·0	15·2	611	193·6
	94	140	684	254	12·4	19·0	15·2	611	178·4
	84	125	678	253	11·7	16·2	15·2	611	159·4
24 × 12	160	238	633	312	18·6	31·4	16·5	532	303·5
	120	179	617	307	14·1	23·6	16·5	532	227·7
	100	149	610	305	11·9	19·7	16·5	532	189·9
24 × 9	94	140	617	230	13·1	22·1	12·7	543	178·2
	84	125	612	229	11·9	19·6	12·7	543	159·4
	76	113	607	228	11·2	17·3	12·7	543	144·3
	68	101	602	227·6	10·6	14·8	12·7	543	129·0
21 × 13	142	211	545	334	16·7	27·8	16·5	450	269·6
	127	189	539	332	16·7	25·0	16·5	450	241·2
	112	167	533	330	13·4	22·0	16·5	450	212·7
21 × 8¼	82	122	545	212	12·8	21·3	12·7	473	155·6
	73	109	539	211	11·6	18·8	12·7	473	138·4
	68	101	537	210	10·9	17·4	12·7	473	129·1
	62	92	533	209	10·2	15·6	12·7	473	117·6
	55	82	528	208·7	9·6	13·2	12·7	473	104·3
18 × 7½	66	98	467	193	11·4	19·6	10·2	404	125·2
	60	89	464	192	10·6	17·7	10·2	404	113·8
	55	82	460	191	9·9	16·0	10·2	404	104·4
	50	74	457	190·5	9·1	14·5	10·2	404	94·9
	45	67	454	190	8·5	12·7	10·2	404	85·4

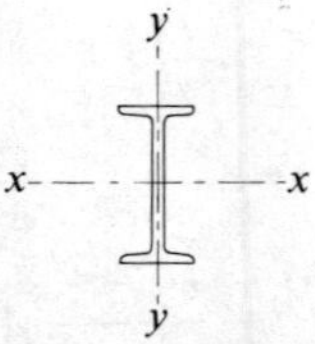

Table 2. 9. 1

Universal Beams

Dimensions and Properties

Moment of Inertia			Radius of Gyration		Elastic Modulus		Plastic Modulus	
Axis x-x		Axis	Axis	Axis	Axis	Axis	Axis	Axis
Gross	Net	y-y	x-x	y-y	x-x	y-y	x-x	y-y
cm^4	cm^4	cm^4	cm	cm	cm^3	cm^3	cm^3	cm^3
717328	649238	42479	38·1	9·27	15586	2021	17628	3206
623868	564909	36250	37·8	9·11	13691	1733	15445	2756
503780	442745	14793	37·0	6·34	10874	961	12566	1552
435802	383016	12512	36·8	6·23	9490	819	10930	1322
375110	330425	10424	36·3	6·05	8241	686	9505	1112
324715	287237	8632	35·6	5·81	7192	569	8345	932
339130	317451	10662	34·3	6·08	7971	726	9144	1166
278833	261485	8385	33·6	5·83	6633	574	7635	929
245412	230504	7111	33·1	5·64	5879	438	6795	797
239463	222705	7701	30·9	5·54	6223	575	7156	925
204747	190651	6377	30·5	5·38	5374	478	6186	773
168535	157219	5002	30·0	5·16	4471	377	5163	615
169843	157215	6225	28·0	5·36	4902	487	5616	781
150015	138896	5391	27·8	5·28	4364	424	4989	680
135973	125997	4789	27·6	5·18	3979	377	4552	608
117700	109262	3992	27·2	5·00	3472	315	3987	513
207252	180191	14973	26·1	7·02	6549	961	7447	1522
151313	131467	10572	25·8	6·81	4901	689	5512	1092
124342	108020	8473	25·6	6·68	4079	556	4562	884
111675	102339	4253	25·0	4·88	3620	370	4141	591
98410	90218	3676	24·8	4·80	3217	321	3672	514
87262	80079	3184	24·6	4·70	2874	279	3283	449
75549	69474	2658	24·2	4·54	2509	234	1877	379
141682	123941	16064	22·9	7·72	5199	963	5849	1518
125619	109835	14093	22·8	7·64	4657	850	5212	1340
109110	94517	12058	22·6	7·53	4091	730	4560	1156
76078	69119	3208	22·1	4·54	2794	303	3198	484
66610	60541	2755	21·9	4·46	2469	262	2820	419
61531	55492	2512	21·8	4·41	2293	239	2616	383
55225	50243	2212	21·7	4·34	2072	211	2362	340
47363	43189	1826	21·3	4·18	1793	175	2051	284
45653	40978	2216	19·1	4·21	1054	230	2229	366
40957	36774	1960	19·0	4·15	1767	204	2012	325
37039	33278	1746	18·8	4·09	1610	183	1830	292
33324	29944	1547	18·7	4·04	1458	162	1654	260
29337	26396	1328	18·5	3·95	1293	140	1469	225

Note: One hole is deducted from each flange under 12 inches wide (serial size) and two holes from each flange 12 inches and over (serial size), in calculating the Net Moment of Inertia about x-x.

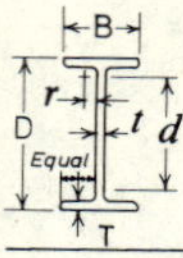

Table 2.9.2
Universal beams
Dimensions and properties

Serial Size	Weight per foot	Weight per metre	Depth of Section D	Width of Section B	Thickness Web t	Thickness Flange T	Root Radius r	Depth between Fillets d	Area of Section
in	lb	kg	mm	mm	mm	mm	mm	mm	cm²
18 × 6	55	82	465	153·5	10·7	18·9	10·2	404	104·4
	50	74	461	152·7	9·9	17·0	10·2	404	94·9
	45	67	457	151·9	9·1	15·0	10·2	404	85·3
16 × 7	50	74	413	180	9·7	16·0	10·2	357	94·9
	45	67	409	179	8·8	14·3	10·2	357	85·4
	40	60	406	178	7·8	12·8	10·2	357	75·9
	36	54	403	177·6	7·6	10·9	10·2	357	68·3
16 × 6	50	74	416	154	10·1	18·1	10·2	357	94·8
	45	67	412	153	9·4	16·0	10·2	357	85·3
	40	59	408	152	8·6	13·9	10·2	357	75·8
16 × 5½	31	46	402	142·4	6·9	11·2	10·2	357	58·9
	26	39	397	141·8	6·3	8·6	10·2	357	49·3
15 × 6	45	67	389	154	9·7	16·3	10·2	333	85·4
	40	60	385	153	8·7	14·4	10·2	333	75·9
	35	52	381	152	7·8	12·4	10·2	333	66·4
14 × 6¾	45	67	364	173	9·1	15·7	10·2	309	85·3
	38	57	359	172	8·0	13·0	10·2	309	72·1
	34	51	356	171·5	7·3	11·5	10·2	309	64·5
	30	45	352	171	6·9	9·7	10·2	309	56·9
14 × 5	26	39	353	126	6·5	10·7	10·2	309	49·3
	22	33	348	125	5·9	8·5	10·2	309	41·7
12 × 6½	36	54	311	167	7·7	13·7	8·9	263	68·3
	31	46	307	166	6·7	11·8	8·9	263	58·8
	27	40	304	165	6·1	10·2	8·9	263	51·4
12 × 5	32	48	310	125	8·9	14·0	8·9	263	60·8
	28	42	307	124	8·0	12·1	8·9	263	53·1
	25	37	304	123·5	7·2	10·7	8·9	263	47·4
12 × 4	22*	33	313	102·4	6·6	10·8	7·6	275	41·8
	19*	28	309	101·9	6·1	8·9	7·6	275	36·3
	16·5*	25	305	101·6	5·8	6·8	7·6	275	31·4
10 × 5¾	29	43	260	147	7·3	12·7	7·6	216	55·0
	25	37	256	146·4	6·4	10·9	7·6	216	47·4
	21	31	251	146	6·1	8·6	7·6	216	39·9
10 × 4	19*	28	260	102·1	6·4	10·0	7·6	224	36·2
	17*	25	257	101·9	6·1	8·4	7·6	224	32·1
	15*	22	254	101·6	5·8	6·8	7·6	224	28·4
8 × 5¼	20	30	207	134	6·3	9·6	7·6	170	38·0
	17	25	203	133	5·8	7·8	7·6	170	32·3

Note: Beams marked * have 2% taper flanges; all other beams have 5% taper flanges.

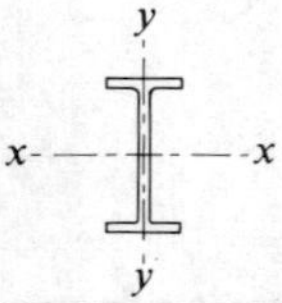

Table 2.9.2
Universal beams
Dimensions and properties

Moment of Inertia			Radius of Gyration		Elastic Molulus		Plastic Modulus	
Axis x-x		Axis	Axis	Axis	Axis	Axis	Axis	Axis
Gross	Net	y-y	x-x	y-y	x-x	y-y	x-x	y-y
cm^4	cm^4	cm^4	cm	cm	cm^3	cm^3	cm^3	cm^3
36160	32275	1093	18·6	3·24	1555	142	1797	229
32380	28917	963	18·5	3·18	1404	126	1620	203
28522	25501	829	18·3	3·12	1248	109	1439	176
27279	24287	1448	17·0	3·91	1322	161	1502	257
24279	21619	1269	16·9	3·86	1186	142	1343	226
21458	19101	1104	16·8	3·81	1056	124	1191	198
18576	16591	922	16·5	3·67	923	104	1046	167
26938	23981	1047	16·9	3·32	1294	136	1486	218
23798	21200	908	16·7	3·26	1155	119	1323	190
20619	18396	786	16·5	3·18	1011	101	1158	162
15603	13838	500	16·3	2·91	776	70·3	886	113
12408	11062	373	15·9	2·75	625	52·6	719	85·8
21276	18949	947	15·8	3·33	1095	123	1254	196
18632	16599	814	15·7	3·27	968	106	1106	170
16046	14300	685	15·5	3·21	842	90·0	959	144
19483	17215	1278	15·1	3·87	1071	148	1210	234
16038	14185	1026	14·9	3·78	894	119	1007	190
14118	12495	885	14·8	3·71	794	103	893	165
12052	10697	730	14·6	3·58	685	85·4	772	137
10054	8765	333	14·3	2·60	570	52·9	652	85·1
8167	7152	257	14·0	2·48	469	41·0	538	66·6
11686	10243	988	13·1	3·81	752	119	843	187
9924	8695	825	13·0	3·73	646	99·5	721	157
8500	7455	691	12·9	3·66	560	83·7	623	133
9485	8215	438	12·5	2·68	611	69·9	705	112
8124	7041	367	12·4	2·63	530	59·0	609	94·9
7143	6195	316	12·3	2·58	470	51·1	539	82·3
6482	5624	189	12·5	2·13	415	37·0	480	59·1
5415	4719	153	12·2	2·05	351	30·0	407	48·2
4381	3851	116	11·8	1·92	287	22·9	337	37·2
6546	5748	633	10·9	3·40	504	86·0	567	135
5544	4866	528	10·8	3·33	433	72·1	484	114
4427	3900	406	10·5	3·18	352	55·5	395	88·8
4004	3457	174	10·5	2·19	308	34·1	353	54·1
3404	2953	144	10·3	2·11	265	28·2	305	45·1
2863	2498	116	10·0	2·02	225	22·8	262	36·8
2880	2493	354	8·71	3·05	379	52·9	313	83·7
2348	2040	280	8·53	2·95	231	41·9	259	67·0

Note: One hole is deducted from each flange in calculating the Net Moment of Inertia about x-x.

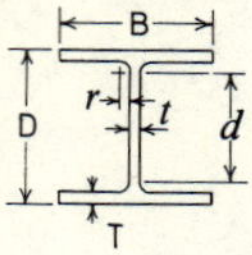

Table 2.9.3
Universal columns
Parallel flanges
Dimensions and properties

Serial Size	Weight per foot	Weight per metre	Depth of Section **D**	Width of Section **B**	Thickness Web **t**	Thickness Flange **T**	Root Radius **r**	Depth between Fillets **d**	Area of Section
in	lb	kg	mm	mm	mm	mm	mm	mm	cm^2
14 × 16	426	634	475	424	47·6	77·0	15·2	290	808·1
	370	550	456	418	42·0	67·5	15·2	290	701·8
	314	467	437	412	35·9	58·0	15·2	290	595·5
	264	393	419	407	30·6	49·2	15·2	290	500·9
	228	339	406	403	26·5	42·9	15·2	290	432·7
	193	287	394	399	22·6	36·5	15·2	290	366·0
	158	235	381	395	18·5	30·2	15·2	290	299·8
Column Core	320	476	427	424	48·0	53·2	15·2	290	607·2
14 × 14½	136	202	375	374	16·8	27·0	15·2	290	257·9
	119	177	368	372	14·5	23·8	15·2	290	225·7
	103	153	362	370	12·6	20·7	15·2	290	195·2
	87	129	356	368	10·7	17·5	15·2	290	164·9
12 × 12	190	283	365	322	26·9	44·1	15·2	247	360·4
	161	240	353	318	23·0	37·7	15·2	247	305·6
	133	198	340	314	19·2	31·4	15·2	247	252·3
	106	158	327	311	15·7	25·0	15·2	247	201·2
	92	137	321	309	13·8	21·7	15·2	247	174·6
	79	117	314	307	11·9	18·7	15·2	247	149·8
	65	97	308	305	9·9	15·4	15·2	247	123·3
10 × 10	112	167	289	265	19·2	31·7	12·7	200	212·4
	89	133	276	261	15·6	25·3	12·7	200	168·9
	72	107	267	258	13·0	20·5	12·7	200	136·6
	60	89	260	256	10·5	17·3	12·7	200	114·0
	49	73	254	254	8·6	14·2	12·7	200	92·9
8 × 8	58	86	222	209	13·0	20·5	10·2	161	110·1
	48	71	216	206	10·3	17·3	10·2	161	91·1
	40	59	210	205	9·3	14·2	10·2	161	75·8
	35	52	206	204	8·0	12·5	10·2	161	66·4
	31	46	203	203	7·3	11·0	10·2	161	58·8
6 × 6	25	37	162	154	8·1	11·5	7·6	123·5	47·4
	20	30	157	153	6·6	9·4	7·6	123·5	38·2
	15·7	23	152	152	6·1	6·8	7·6	123·5	29·8

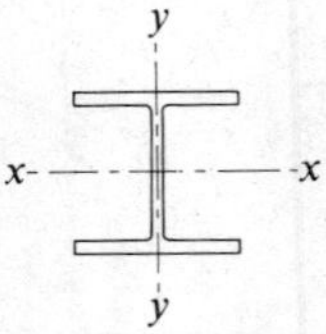

Table 2.9.3
Universal columns
Parallel flanges
Dimensions and properties

Moment of Inertia			Radius of Gyration		Elastic Modulus		Plastic Modulus	
Axis x-x		Axis	Axis	Axis	Axis	Axis	Axis	Axis
Gros	Net	y-y	x-x	y-y	x-x	y-y	x-x	y-y
cm^4	cm^4	cm^4	cm	cm	cm^3	cm^3	cm^3	cm^3
275140	245765	98211	18·4	11·0	11592	4632	14247	7114
227023	202556	82665	18·0	10·8	9964	3951	12078	6058
183119	163167	67905	17·5	10·7	8388	3283	10009	5038
146765	130632	55410	17·1	10·5	7004	2723	8229	4157
122474	108919	46816	16·8	10·4	6027	2324	6994	3541
99994	88860	38714	16·5	10·3	5080	1940	5818	2952
79110	70243	31008	16·3	10·2	4153	1570	4689	2384
172391	154580	68057	16·8	10·6	8075	3207	9700	4979
66307	58518	23632	16·0	9·58	3540	1262	3976	1917
57133	50410	20470	15·9	9·53	3104	1100	3457	1668
48525	42789	17470	15·8	9·45	2681	944	2964	1430
40246	35484	14555	15·6	9·40	2264	790	2482	1196
78777	67879	24545	14·8	8·26	4314	1525	5101	2337
64177	55230	20239	14·5	8·13	3641	1273	4245	1947
50832	43696	16230	14·2	8·03	2991	1034	3436	1576
38740	33282	12524	13·9	7·90	2368	806	2680	1228
32838	28204	10673	13·7	7·82	2049	691	2298	1052
27601	23700	9006	13·6	7·75	1756	587	1953	892
22202	19063	7268	13·4	7·67	1442	477	1589	723
29914	27401	9796	11·9	6·78	2070	741	2417	1132
22575	20670	7519	11·6	6·68	1634	576	1875	879
17510	16025	5901	11·3	6·58	1313	457	1485	695
14307	13082	4849	11·2	6·53	1099	379	1228	575
11360	10389	3873	11·05	6·45	895	305	989	462
9462	8464	3119	9·27	5·33	851	299	979	456
7647	6830	2536	9·17	5·28	708	246	802	374
6088	5444	2041	8·97	5·18	581	199	652	303
5263	4703	1770	8·89	5·16	510	174	568	264
4564	4083	1539	8·81	5·11	449	151	497	230
2218	1948	709	6·83	3·86	274	91·8	310	140
1742	1532	558	6·76	3·81	221	73·1	247	111
1263	1113	403	6·50	3·68	166	52·9	184	80·9

Note: One hole is deducted from each flange under 12 inches wide and two holes from each flange 12 inches and over, in calculating the Net Moment about x-x.

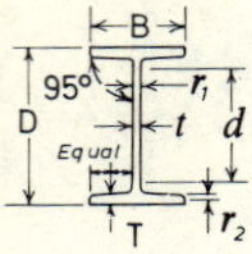

Table 2.9.4

Joists

Dimensions and properties

Size D	B	Weight per foot	Weight per metre	Thickness Web t	Thickness Flange T	Radius Root r_1	Radius Toe r_2	Depth between Fillets d	Area of Section
in	mm	lb	kg	mm	mm	mm	mm	mm	cm²
8 × 4	**203 × 102**	17	25·30	5·8	10·4	9·4	3·2	161·1	32·26
7 × 4	**178 × 102**	14·5	21·56	5·3	9·1	9·4	3·2	138·2	27·49
6 × 3½	**152 × 89**	11·5	17·10	4·9	8·3	7·9	2·4	117·7	21·81
5 × 3	**127 × 76**	9	13·38	4·5	7·6	7·9	2·4	94·2	17·06
4 × 2½	**102 × 63**	6·5	9·66	4·1	6·6	6·9	2·4	73·2	12·32
3 × 2	**76 × 51**	4·5	6·70	3·8	5·7	6·9	2·4	50·2	8·54

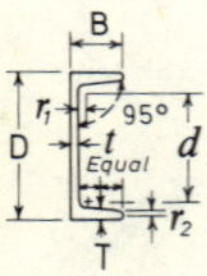

Table 2.9.5

Channels

Dimensions and properties

Size D × B		Weight per foot	Weight per metre	Thickness Web t	Thickness Flange T	Radius Root r_1	Radius Toe r_2	Depth between Fillets d	Area of Section
in	mm	lb	kg	mm	mm	mm	mm	mm	cm²
17 × 4	**432 × 102**	44	65·48	12·2	16·8	15·2	4·8	362·4	83·49
15 × 4	**381 × 102**	37	55·05	10·4	16·3	15·2	4·8	312·4	70·19
12 × 4	**305 × 102**	31	46·14	10·2	14·8	15·2	4·8	239·2	58·83
12 × 3½	**305 × 89**	28	41·65	10·2	13·7	13·7	3·2	245·4	53·11
10 × 3½	**254 × 89**	24	35·70	9·1	13·6	13·7	3·2	194·7	45·52
10 × 3	**254 × 76**	19	28·26	8·1	10·9	12·2	3·2	203·8	36·03
9 × 3½	**229 × 89**	22	32·73	8·6	13·3	13·7	3·2	169·8	41·73
9 × 3	**229 × 76**	17·5	26·04	7·6	11·2	12·2	3·2	178·0	33·20
8 × 3½	**203 × 89**	20	29·76	8·1	12·9	13·7	3·2	145·2	37·94
8 × 3	**203 × 76**	16	23·80	7·1	11·2	12·2	3·2	152·5	20·94
7 × 3½	**178 × 89**	18	26·79	7·6	12·3	13·7	3·2	121·0	34·15
7 × 3	**178 × 76**	14	20·82	6·6	10·3	12·2	3·2	128·8	16·54
6 × 3½	**152 × 89**	16	23·81	7·1	11·6	13·7	3·2	97·0	30·36
6 × 3	**152 × 76**	12	17·86	6·3	9·0	12·2	2·4	105·9	22·77
5 × 2½	**127 × 63**	10	14·89	6·3	9·2	10·7	2·4	84·0	18·98
4 × 2	**102 × 51**	7	10·41	6·1	7·6	9·1	2·4	65·7	13·28
3 × 1½	**76 × 58**	4·5	6·69	5·1	6·8	7·6	2·4	45·8	8·53

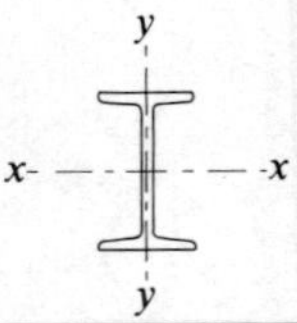

Table 2.9.4

Joists

Dimensions and properties

Size	Moment of Inertia			Radius of Gyration		Elastic Modulus		Plastic Modulus	
	Axis x-x								
D × B	Gross	Net	Axis y-y	Axis x-x	Axis y-y	Axis x-x	Axis y-y	Axis x-x	Axis y-y
in	cm^4	cm^4	cm^4	cm	cm	cm^3	cm^3	cm^3	cm^3
8 × 4	2294	1957	162·6	8·43	2·25	225·8	32·02	256·3	51·78
7 × 4	1522	1297	139·7	7·44	2·25	171·2	27·50	193·4	44·61
6 × 3½	883·1	760·1	86·29	6·36	1·99	115·9	19·41	131·2	31·39
5 × 3	477·0	399·3	50·37	5·29	1·72	75·12	13·22	85·44	21·37
4 × 2½	218·2	184·9	25·41	4·21	1·44	42·95	8·00	49·11	12·96
3 × 2	83·07	69·57	11·22	3·12	1·15	21·80	4·42	25·22	7·21

Note: One hole is deducted from each flange in calculating the Net Moment of Inertia about x-x.

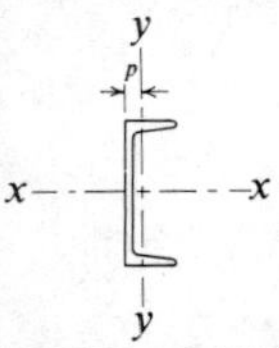

Table 2.9.5

Channels

Dimensions and properties

Size	Dimension	Moment of Inertia			Radius of Gyration		Elastic Modulus	
		Axis x-x						
D × B	p	Gross	Net	Axis y-y	Axis x-x	Axis y-y	Axis x-x	Axis y-y
inches	cm	cm^4	cm^4	cm^4	cm	cm	cm^3	cm^3
17 × 4	2·32	21399	18405	628·6	16·01	2·74	991·1	80·15
15 × 4	2·52	14894	12651	579·8	14·57	2·87	781·8	75·87
12 × 4	2·66	8214	6926	499·6	11·82	2·91	539·0	66·60
12 × 3½	2·18	7061	5865	325·4	11·53	2·47	463·3	48·49
10 × 3½	2·42	4448	3636	302·4	9·88	2·58	350·2	46·71
10 × 3	1·86	3367	2699	162·6	9·67	2·12	265·1	28·22
9 × 3½	2·53	3387	2750	285·0	9·01	2·61	296·4	44·82
9 × 3	2·00	2610	2065	158·7	8·87	2·19	228·3	28·22
8 × 3½	2·65	2491	2009	264·4	8·10	2·64	245·2	42·34
8 × 3	2·13	1950	1525	151·4	8·02	2·23	192·0	27·59
7 × 3½	2·76	1753	1405	241·0	7·16	2·66	197·2	39·29
7 × 3	2·20	1337	1039	134·0	7·10	2·25	150·4	24·73
6 × 3½	2·86	1166	928·7	215·1	6·19	2·66	153·0	35·70
6 × 3	2·21	851·6	660·0	113·8	6·11	2·24	111·8	21·05
5 × 2½	1·94	482·6	350·3	67·24	5·04	1·88	75·99	15·25
4 × 2	1·51	207·7	148·8	29·11	3·95	1·48	40·89	8·16
3 × 1½	1·19	74·14	50·73	10·66	2·95	1·12	19·46	4·08

One hole is deducted from each flange in calculating the Net Moment of Inertia about x-x.

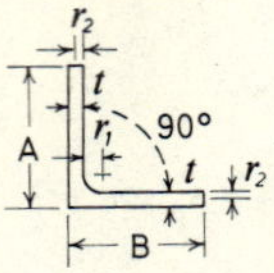

Table 2.9.6

Equal Angles

Dimensions and Properties

Size A × B in	Nominal Thickness in	Size A × B mm	Thickness t (actual) mm	Calculated Weight per metre kg	Sectional Area cm^2
8 × 8	7/8	**203·2 × 203·2**	22.12	66.99	85.42
	3/4		18·95	57·89	73·82
	5/8		15·77	48·64	62·02
6 × 6	3/4	**152·4 × 152·4**	18·97	42·71	54·45
	5/8		15·82	36·04	45·95
	1/2		12·60	29·04	37·03
	3/8		9·42	22·00	28·06
5 × 5	5/8	**127·0 × 127·0**	15·80	29·63	37·78
	1/2		12·60	23·97	30·56
	3/8		9·47	18·28	23·31
4 × 4	5/8	**101·6 × 101·6**	15·85	23·35	29·78
	1/2		12·60	18·90	24·09
	3/8		9·45	14·42	18·39
3½ × 3½	1/2	**88·9 × 88·9**	12·60	16·36	20·87
	3/8		9·42	12·49	15·92
	1/4		6·27	8·48	10·81
3 × 3	1/2	**76·2 × 76·2**	12·60	13·83	17·64
	3/8		9·40	10·56	13·47
	1/4		6·22	7·16	9·12
2½ × 2½	3/8	**63·5 × 63·5**	9·45	8·77	11·18
	5/16		7·90	7·44	9·48
	1/4		6·22	5·95	7·59
2¼ × 2¼	5/16	**57·2 × 57·2**	7·77	6·55	8·35
	1/4		6·25	5·35	6·82
	3/16		4·60	4·01	5·11
2 × 2	5/16	**50·8 × 50·8**	7·82	5·80	7·39
	1/4		6·32	4·77	6·08
	3/16		4·65	3·58	4·56

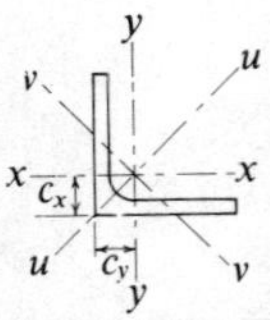

Table 2. 9. 6

Equal Angles

Dimensions and Properties

Centre of Gravity		Moment of Inertia				Radius of Gyration				Elastic Modulus	
Cx	Cy	Axis x-x	Axis y-y	Axis u-u Max.	Axis v-v Min.	Axis x-x	Axis y-y	Axis u-u Max.	Axis v-v Min.	Axis x-x	Axis y-y
cm	cm	cm⁴	cm⁴	cm⁴	cm⁴	cm	cm	cm	cm	cm³	cm³
5.87	5·87	3294	3294	5232	1357	6·21	6·21	7·83	3·99	228	228
5·75	5·75	2885	2885	4587	1183	6·25	6·25	7·88	4·00	198	198
5·63	5·63	2455	2455	3907	1004	6·29	6·29	7·94	4·02	167	167
4·49	4·49	1164	1164	1846	482	4·62	4·62	5·82	2·98	108	108
4·37	4·37	999	999	1587	411	4·66	4·66	5·88	2·99	91·9	91·9
4·24	4·24	819	819	1303	335	4·70	4·70	5·93	3·01	74·5	74·5
4·11	4·11	631	631	1003	258	4·74	4·74	5·98	3·03	56·7	56·7
3·73	3·73	560	560	888	232	3·85	3·85	4·85	2·48	62·4	62·4
3·61	3·61	462	462	734	190	3·89	3·89	4·90	2·49	50·8	50·8
3·49	3·49	359	359	571	147	3·92	3·92	4·95	2·51	39·0	39·0
3·10	3·10	275	275	434	116	3·04	3·04	3·82	1·97	38·9	38·9
2·98	2·98	228	228	361	94·3	3·08	3·08	3·87	1·98	31·7	31·7
2·86	2·86	178	178	283	73·1	3·11	3·11	3·92	1·99	24·4	24·4
2·66	2·66	149	149	235	62·0	2·67	2·67	3·36	1·72	23·9	23·9
2·54	2·54	116	116	185	47·9	2·70	2·70	3·41	1·74	18·3	18·3
2·41	2·41	81·0	81·0	129	33·3	2·74	2·74	3·45	1·75	12·5	12·5
2·35	2·35	90·4	90·4	143	38·2	2·26	2·26	2·84	1·47	17·1	17·1
2·23	2·23	71·1	71·1	113	29·5	2·30	2·30	2·89	1·48	13·2	13·2
2·10	2·10	49·6	49·6	78·8	20·3	2·33	2·33	2·94	1·49	8·97	8·97
1·92	1·92	40·5	40·5	64·0	17·0	1·90	1·90	2·38	1·23	9·15	9·15
1·86	1·86	35·0	35·0	55·5	14·6	1·92	1·92	2·42	1·24	7·80	7·80
1·80	1·80	28·6	28·6	45·4	11·8	1·94	1·94	2·45	1·25	6·28	6·28
1·70	1·70	24·7	24·7	39·1	10·3	1·72	1·72	2·16	1·11	6·15	6·15
1·64	1·64	20·6	20·6	32·6	8·53	1·74	1·74	2·19	1·12	5·05	5·05
1·57	1·57	15·8	15·8	25·0	6·51	1·76	1·76	2·21	1·13	3·81	3·81
1·54	1·54	17·0	17·0	26·8	7·17	1·52	1·52	1·91	·98	4·81	4·81
1·49	1·49	14·3	14·3	22·7	5·95	1·53	1·53	1·93	·99	3·98	3·98
1·42	1·42	11·0	11·0	17·4	4·54	1·55	1·55	1·95	1·00	3·00	3·00

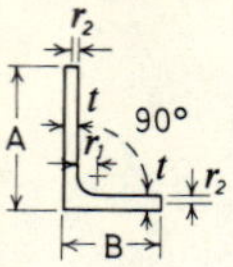

Table 2.9.7

Unequal angles

Dimensions and properties

Metric

Size A × B	Nominal Thickness	Size A × B	Thickness t (actual)	Calculated Weight per metre	Sectional Area
in	in	mm	mm	kg	cm²
9 × 4	¾ ½	**228·6 × 101·6**	18·92 12·57	46·40 31·53	59·17 40·20
8 × 6	¾ ½	**203·2 × 152·4**	18·95 12·57	50·27 34·06	64·10 43·44
8 × 4	⅝ ½	**203·2 × 101·6**	15·82 12·60	36·04 29·04	45·95 37·03
7 × 3½	½ 7/16	**177·8 × 88·9**	12·62 11·07	25·29 22·34	32·24 28·48
6 × 4	⅝ ½ ⅜	**152·4 × 101·6**	15·80 12·60 9·47	29·63 23·97 18·28	37·78 30·56 23·31
6 × 3½	½ ⅜	**152·4 × 88·9**	12·62 9·42	22·75 17·24	29·00 21·99
6 × 3	½ ⅜	**152·4 × 76·2**	12·60 9·47	21·43 16·37	27·33 20·87
5 × 3½	½ ⅜	**127·0 × 88·9**	12·65 9·42	20·24 15·34	25·81 19·56
5 × 3	½ ⅜ 5/16	**127 × 76·2**	12·60 9·45 7·82	18·90 14·42 12·05	24·09 18·39 15·37
4 × 3½	½ ⅜ 5/16	**101·6 × 88·9**	12·65 9·50 7·85	17·70 13·54 11·30	22·57 17·27 14·41

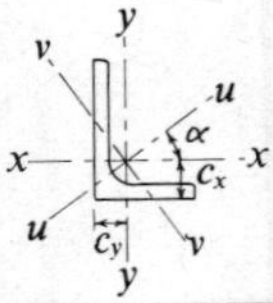

Table 2.9.7
Unequal angles
Dimensions and properties

Metric

Centre of Gravity		Moment of Inertia				Radius of Gyration					Elastic Modulus	
Cx	Cy	Axis x-x	Axis y-y	Axis u-u Max.	Axis v-v Min.	Axis x-x	Axis y-y	Axis u-u Max.	Axis v-v Min.	Angle Axis u-u	Axis x-x	Axis y-y
cm	cm	cm^4	cm^4	cm^4	cm^4	cm	cm	cm	cm	tan α	cm^3	cm^3
8·60	2·29	3154	396	3283	267	7·30	2·59	7·45	2·13	·21	221	50·3
8·33	2·04	2197	283	2292	187	7·39	2·65	7·55	2·16	·22	151	34·8
6·47	3·95	2625	1268	3209	684	6·40	4·45	7·08	3·27	·55	190	112
6·22	3·70	1826	890	2240	476	6·48	4·53	7·18	3·31	·55	129	77·1
7·33	2·29	1947	333	2064	216	6·51	2·69	6·70	2·17	·26	150	42·4
7·20	2·16	1591	276	1689	177	6·55	2·73	6·75	2·19	·26	121	34·5
6·36	1·95	1051	181	1115	117	5·71	2·37	5·88	1·90	·26	92·0	26·0
6·29	1·89	935	162	993	104	5·73	2·38	5·91	1·91	·26	81·4	23·1
5·13	2·61	871	309	1000	180	4·80	2·86	5·14	2·19	·43	86·1	41·0
5·00	2·48	716	257	825	148	4·84	2·90	5·19	2·20	·44	70·0	33·4
4·88	2·36	555	201	641	115	4·88	2·93	5·24	2·22	·44	53·6	25·7
5·25	2·10	685	174	752	107	4·86	2·45	5·09	1·92	·34	68·5	25·6
5·11	1·97	528	136	581	82·6	4·90	2·48	5·14	1·94	·35	52·1	19·6
5·52	1·74	647	110	685	71·5	4·87	2·01	5·01	1·62	·26	66·6	18·7
5·39	1·62	503	867	534	55·7	4·91	2·04	5·06	1·63	·26	51·0	14·4
4·17	2·28	412	166	484	93·9	3·99	2·54	4·33	1·91	·48	48·3	25·1
4·04	2·16	318	129	375	72·5	4·03	2·57	4·38	1·92	·48	36·8	19·2
4·41	1·89	389	105	429	64·4	4·02	2·09	4·22	1·63	·35	46·9	18·3
4·28	1·77	302	82·8	335	50·0	4·06	2·12	4·27	1·65	·36	35·9	14·1
4·21	1·70	255	70·2	283	42·3	4·07	2·14	4·29	1·66	·36	30·1	11·9
3·15	2·52	219	156	299	75·0	3·11	2·62	3·64	1·82	·75	31·2	24·4
3·03	2·40	171	122	235	58·2	3·15	2·66	3·69	1·84	·75	24·0	18·8
2·96	2·33	145	103	199	49·1	3·17	2·68	3·72	1·85	·75	20·1	15·8

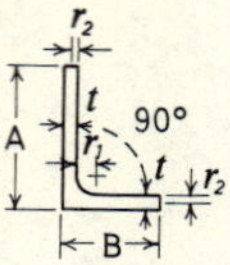

Table 2.9.8

Unequal angles

Dimensions and properties
Metric

Size A × B	Nominal Thickness	Size A × B	Thickness t (actual)	Calculated Weight per metre	Sectional Area
in	in	mm	mm	kg	cm^2
4 × 3	½	**101·6 × 76·2**	12·60	16·36	20·87
	⅜		9·42	12·49	15·92
	5/16		7·90	10·57	13·48
4 × 2½	⅜	**101·6 × 63·5**	9·47	11·60	14·79
	5/16		7·82	9·68	12·35
	¼		6·30	7·88	10·05
3½ × 3	½	**88·9 × 76·2**	12·67	15·19	19·36
	⅜		9·47	11·60	14·79
	5/16		7·82	9·68	12·35
3½ × 2½	⅜	**88·9 × 63·5**	9·40	10·56	13·47
	5/16		7·85	8·92	11·37
	¼		6·22	7·16	9·12
3 × 2½	⅜	**76·2 × 63·5**	9·45	9·67	12·33
	5/16		7·90	8·18	10·43
	¼		6·25	6·55	8·36
3 × 2	⅜	**76·2 × 50·8**	9·45	8·77	11·18
	5/16		7·90	7·44	9·48
	¼		6·22	5·95	7·59
2½ × 2	5/16	**63·5 × 50·8**	7·77	6·55	8·35
	¼		6·25	5·35	6·82
	3/16		4·60	4·01	5·11
2½ × 1½	¼	**63·5 × 38·1**	6·32	4·77	6·08
	3/16		4·65	3·58	4·56
2 × 1½	¼	**50·8 × 38·1**	6·12	4·01	5·12
	3/16		4·67	3·13	3·99

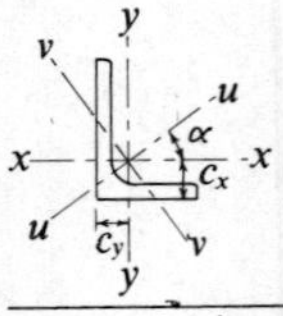

Table 2. 9. 8

Unequal angles

Dimensions and properties

Metric

Centre of gravity		Moment of Inertia				Radius of Gyration					Elastic Modulus	
Cx	Cy	Axis **x-x**	Axis **y-y**	Axis u-u Max.	Axis v-v Min.	Axis **x-x**	Axis **y-y**	Axis u-u Max.	Axis v-v Min.	Angle Axis **u-u**	Axis **x-x**	Axis **y-y**
cm	cm	cm^4	cm^4	cm^4	cm^4	cm	cm	cm	cm	tan α	cm^3	cm^3
3·34	2·08	207	98·8	251	54·3	3·15	2·18	3·47	1·61	·54	30·3	17·8
3·22	1·96	162	77·8	197	42·0	3·19	2·21	3·52	1·62	·55	23·3	13·7
3·16	1·90	138	66·8	169	35·9	3·20	2·23	3·54	1·63	·55	19·7	11·7
3·45	1·56	151	45·8	171	27·5	3·21	1·76	3·40	1·36	·38	22·7	9·56
3·38	1·49	129	38·9	145	23·2	3·23	1·78	3·42	1·37	·39	19·0	8·02
3·31	1·43	106	32·2	119	19·1	3·25	1·79	3·44	1·38	·39	15·4	6·54
2·84	2·21	142	95·4	190	47·5	2·71	2·22	3·13	1·57	·71	23·4	17·6
2·71	2·08	111	75·1	150	36·7	2·74	2·25	3·18	1·58	·72	18·0	13·6
2·65	2·02	94·2	63·7	127	30·9	2·76	2·27	3·21	1·58	·72	15·1	11·4
2·91	1·65	104	43·9	123	24·6	2·78	1·80	3·02	1·35	·49	17·4	9·34
2·85	1·59	88·8	37·7	106	21·0	2·79	1·82	3·05	1·36	·50	14·7	7·92
2·78	1·53	72·1	30·7	85·8	17·0	2·81	1·83	3·07	1·37	·50	11·8	6·37
2·40	1·77	67·3	42·1	88·2	21·3	2·34	1·85	2·67	1·31	·67	12·9	9·20
2·34	1·71	57·8	36·2	75·9	18·1	2·35	1·86	2·70	1·32	·68	10·9	7·81
2·27	1·64	46·9	29·5	61·7	14·7	2·37	1·88	2·72	1·33	·68	8·77	6·27
2·62	1·36	63·2	22·3	72·3	13·2	2·38	1·41	2·54	1·09	·43	12·6	5·99
2·56	1·30	54·5	19·1	62·5	11·3	2·40	1·43	2·57	1·09	·43	10·8	5·12
2·49	1·24	44·4	15·8	51·1	9·21	2·42	1·45	2·59	1·10	·44	8·66	4·13
2·03	1·40	32·1	18·1	40·7	9·50	1·96	1·47	2·21	1·07	·62	7·44	4·93
1·97	1·35	26·7	15·1	34·0	7·86	1·98	1·49	2·23	1·07	·62	6·10	4·06
1·90	1·28	20·4	11·6	26·0	6·01	2·00	1·51	2·26	1·08	·62	4·59	3·06
2·20	0·94	24·4	6·59	27·0	4·05	2·00	1·04	2·11	0·82	·35	5·89	2·30
2·13	0·88	18·7	5·10	20·7	3·10	2·03	1·06	2·13	0·82	·36	4·43	1·74
1·66	1·03	12·7	6·04	15·4	3·32	1·57	1·09	1·73	0·81	·54	3·70	2·17
1·60	0·98	10·1	4·83	12·3	2·63	1·59	1·10	1·75	0·81	·54	2·89	1·70

Section 3

Reinforced Concrete

Design Requirements
Design of Beams with Tensile Reinforcement Only
Slab Design
Design of Rectangular Beams with Tensile and
    Compressive Steel
Design of T-beams and L-beams
Design of Hollow Tile Floors
Simple Column Bases

3. Reinforced Concrete

3.1 DESIGN REQUIREMENTS

From CP 114: Part 2: 1969 'The structural use of reinforced concrete in buildings.'

Symbols

A_b = equivalent area of helical reinforcement (volume of length of column)
A_c = cross-sectional area of concrete, excluding any finishing material and reinforcing steel
A_k = cross-sectional area of concrete in column core, excluding area of longitudinal reinforcement
A_{so} = cross-sectional area of steel in compression
A_{st} = cross-sectional area of steel in tension
A_w = cross-sectional area of stirrup
b = breadth of rectangular beam or breadth of flange of a T-beam or L-beam
b_r = breadth of the rib of a T-beam or L-beam
D = diameter generally
d = overall depth
d_1 = effective depth to the tensile reinforcement in a beam
d_2 = depth to the compressive reinforcement in a beam
d_n = depth of concrete in compression in a beam
d_s = depth of slab forming the flange of a T-beam or L-beam
e = eccentricity of a load on a column
f_y = guaranteed yield or proof stress
K_b = stiffness of beam
K_{b1} = stiffness of beam on one side of a column
K_{b2} = stiffness of beam on the opposite side of a column
K_1 = stiffness of lower column
K_u = stiffness of upper column

L = length of a column or beam between adequate lateral restraints (in flat slabs, L is the average of L_1 and L_2)
L_1 = (in flat slabs) length of panel in the directionof span
L_2 = (in flat slabs) width of panel at right angles to direction of span
l = effective span of beam or slab or effective height of column
l_a = lever arm of the resistance moment
l_x = length of shorter side of slab spanning in two directions
l_y = length of long side of slab spanning in two directions
M = bending moment (suffixes as required)
M_e = bending moment at end of beam framing into a column, assuming fixity at both ends of the beam
M_{es} = the maximum difference between the moments at the ends of two beams framing into opposite sides of a column, each calculated on the assumption that the ends of the beams are fixed and that one of the beams is not loaded
M_r = moment of resistance of a section to bending
M_x and
M_y = maximum bending moments, for spans l_x and l_y respectively, on strips of unit width in slabs spanning in two directions
m = modular ratio
o = sum of perimeters of the bars in the tensile reinforcement
P = permissible load on a short column subject to both direct load and bending
P_b = load on a column as defined by equation 17
P_o = axial load permissible on a short column
p_{oo} = permissible stress in concrete in direct compression
p_{ob} = permissible compressive stress in concrete in bending
p_{so} = permissible compressive stress in the reinforcement
p_{st} = permissible tensile stress in the reinforcement

Q = total shear across a section
q = shear stress at a section of a beam or slab
R = modulus of rupture
s = spacing or pitch of stirrups
u = cube crushing strength of concrete
u_p = u for preliminary test
u_w = u for works test
W = total load on beam or slab
W_d = dead load
W_s = superimposed load
w = total load per unit area of slab or per unit length of beam
α_x and
α_y = bending moment coefficients for the short and long spans respectively, for slabs spanning in two directions and simply supported on four sides
β_x and
β_y = bending moment coefficients for the short and long spans respectively, for rectangular panels supported on four sides and with provision for torsion at corners

Nominal concrete mixes

From Table 6, CP 114. Permissible stresses for nominal concrete mixes with Portland Cement or Portland-Blastfurnace Cement and with aggregates complying with BS 882 or BS 1047 are shown in Table 3. 1. 1.

Modifications to these stresses are allowed for variations in concrete strength, age at loading, wind forces, use of high bond bars and for slender beams. These permissible stresses are based on concretes complying with Table 1 of CP 114, which among other things, requires cube strengths in works tests at 28 days as shown in Table 3. 1. 2.

Table 3.1.1

	Permissible concrete stresses				
	Compressive			Bond	
Mix proportions	Direct N/mm^2	Due to bending N/mm^2	Shear N/mm^2	Average N/mm^2	Local N/mm^2
1 : 1 : 2	7·6	10	0·9	1·00	1·50
1 : 1½ : 3	6·5	8·5	0·8	0·93	1·40
1 : 2 : 4	5·3	7	0·7	0·83	1·25

Table 3.1.2

Mix proportions	Cube strength N/mm^2
1 : 1 : 2	30
1 : 1½ : 3	25·5
1 : 2 : 4	21

Designed concrete mixes

Permissible stresses for designed concrete mixes (from Tables 8 and 10 in CP 114) in N/mm^2.

Table 3.1.3

Compressive stresses			Bond stresses	
Direct	Bending	Shear stress	Average	Local
$0{\cdot}75p_{cb}$	p_{cb}	$\frac{u_w}{50} + 0{\cdot}27$ but $\not> 0{\cdot}90$	$\frac{u_w}{50} + 0{\cdot}40$ but $\not> 1{\cdot}0$	$\frac{2u_w}{75} + 0{\cdot}67$ but $\not> 1{\cdot}5$

u_w = Specified works cube strength at 28 days

$$p_{cb} = \frac{u_w}{2{\cdot}73}$$

As an alternative to using nominal mixes, the concrete mix may be designed within the limitations laid down in Table 4 of CP 114. For O.P.C. concrete, these are from 240 to 540 kg of cement per cubic metre of finished concrete, and values of u_w from 15 to 50 N/mm^2 at 28 days.

The Code requires that the mix should be designed to have a mean strength that exceeds the specified works cube strength by twice the standard deviation. The value of the standard deviation must be calculated from at least 40 cube results, and a value less than 3·5 N/mm^2 cannot be used.

Permissible stresses in steel reinforcement

From Table 11, CP 114

Table 3.1.4

Type of stress	Permissible stress (N/mm^2) M.S. Bars Diameter $\not>$ 40 mm	Diameter $>$ 40 mm	All bars with a guaranteed proof or yield stress (f_y)
Tensile stress other than shear	140	125	$0{\cdot}55f_y$ but $\not>$ 230 if dia. $<$ 20 mm, $\not>$ 210 if dia. $>$ 20 mm
Tensile stress in shear	140	125	$0{\cdot}55f_y$, but $\not>$ 175
Compressive stress	125	110	$0{\cdot}55f_y$ but $\not>$ 175

Cover to reinforcement
From Clause 307, CP 114

Table 3.1.5

	Situation of reinforcement	Cover not less than
1	End of bar	25 mm or twice bar diameter
2	Longitudinal barin column	40 mm or bar diameter
3	As 2 but column less than 200 mm minimum dimension and bars less than 12 mm diameter	25 mm
4	Longitudinal bar in beam	25 mm or bar diameter
5	Slab reinforcement	15 mm or bar diameter
6	Any other reinforcement	15 mm or bar diameter
	External work or corrosive conditions	40 mm overall steel

Distance between bars

Clause 308, CP 114

The horizontal distance between parallel bars should not be less than the greater of:

1. The diameter of equal bars,
2. The diameter of the larger bar,
3. 5 mm more than the nominal maximum size of coarse aggregate.

Stiffness of members

Members should be sufficiently stiff to prevent deflection that will impair the efficiency of the structure or crack finishes or partitions. If steel stresses do not exceed 140 N/mm^2 and concrete stresses do not exceed 10 N/mm^2, stiffness will be satisfactory if Table 3.1.6 (from Table 13, CP 114) below is adhered to. If steel stresses exceed 140 N/mm^2 **or** concrete stresses exceed 10 N/mm^2 90% of Table 3.1.6 values should not be exceeded; if steel stresses exceed 140 N/mm^2 **and** concrete stresses exceed 10 N/mm^2 85% of Table 3.1.6 values should not be exceeded.

Table 3.1.6

	Ratio of span to overall depth
Beams	
Simply supported beams	20
Continuous beams	25
Cantilever beams	10
Slabs	
Slabs spanning in one direction, simply supported	30
Slabs spanning in one direction, continuous	35
Slabs spanning in two directions, simply supported	35
Slabs spanning in two directions, continuous	40
Cantilever slabs	12

Bond and anchorage

Clause 310, CP 114.

1. Bars in tension should extend beyond any section so that average bond stress does not exceed that permitted in Tables 6 and 10, CP 114.

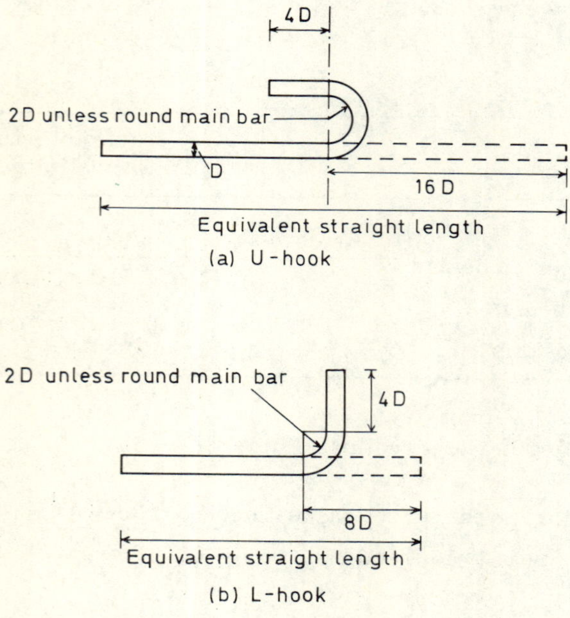

Figure 3.1.1 Types of hook for mild steel bars

This is satisfied if length from section is not less than:

$$\text{Bar diameter} \times \frac{\text{Tensile stress in bar}}{\text{Four times permissible average bond stress}}$$

The bar should extend at least 12 diameters beyond the point at which it is no longer required to resist stress. A length equal to the value of a hook may be deducted from the required length.

2. Local bond stress $= \frac{Q}{l_a o}$ should not exceed the permissible local bond stress given in Tables 3.1.1 and 3.1.3.

3. Bars in compression should extend beyond any section so that average bond stress does not exceed the permitted average bond stress, given in Tables 3.1.1 and 3.1.3 by more than 25%. This is satisfied if the length measured from such a section is not less than

$$\text{Bar diameter} \times \frac{\text{Compressive stress in bar}}{\text{Five times permissible average bond stress}}$$

The bar should extend at least 12 bar diameters beyond the point at which it is no longer required to resist stress.

Laps in bars

1. Bars in tension should have a length of lap not less than:

$$\text{Bar diameter} \times \frac{\text{Tensile stress in bar}}{\text{Four times the permissible average bond stress}}$$

or 30 bar diameters, whichever is the greater.

2. Bars in compression should have a length of lap not less than:

$$\text{Bar diameter} \times \frac{\text{Compressive stress in bar}}{\text{Five times the permissible average bond stress}}$$

or 24 bar diameters, whichever is the greater.

Resistance to shear

Clause 316, CP 114.

1. Shear stress $q = \dfrac{Q}{b\, l_a}$

2. If q is greater than p_q the whole shearing force should be provided by shear reinforcement. In any case, q should not be greater than four times p_q.

3. If q is less than p_q, nominal shear reinforcement should be provided. The spacing should not exceed the lever arm and its area should not be less than 0·15% of the concrete area, calculated on an assumed breadth of beam or rib $6\dfrac{q}{p_q}$.

4. If q does not exceed p_q, no shear reinforcement need be provided in slabs, footings, bases, pile caps and similar members. A conservative approach is recommended in designing members with no shear reinforcement.

5. Stirrups have a shear resistance:

$$\phi = \frac{p_{st}\, A_w\, l_a}{s}$$

6. An inclined bar can be considered as forming the tension member of a lattice girder, the concrete providing the compression member.

Shear resistance of bar $\phi = p_{st} A_{st} \sin \theta$, where θ = angle of inclination of bar to horizontal. If $\theta = 45°$, $\phi = 0{\cdot}707\ p_{st} A_{st}$. If $\theta = 60°$, $\phi = 0{\cdot}5\ p_{st} A_{st}$

Columns

Clause 321, CP 114: Reinforcement.

1. The cross-sectional area of longitudinal reinforcement should not be less than 0·8 percent nor more than 8 percent of the gross cross sectional area of the column. If bars from the column below are to be lapped with those in the column, the reinforcement should not exceed 4%. The bars should not be less than 12mm in diameter.

2. The pitch of transverse reinforcement should not be greater than the least of:

   (a) The least lateral dimension of the column,
   (b) 12 times the diameter of the smallest longitudinal reinforcement,
   (c) 300mm.

   The diameter of the transverse reinforcement should not be less than one quarter of the diameter of the longitudinal reinforcement, nor less than 5mm.

Clause 322a, CP 114: Permissible loads on axially loaded columns.

1. A short column has a ratio of effective length to least lateral dimension which is less than 15. Maximum stresses are those specified in Tables 3.1.1 and 3.1.3.

2. Short columns with lateral ties can carry a maximum axial load:

$$P_o = p_{cc} A_c + p_{sc} A_{sc}$$

3. Short columns with helical reinforcement can carry a maximum axial load equal to the lesser of

$$P_o = p_{cc} A_c + p_{sc} A_{sc}$$

or $$P_o = p_{cc} A_k + p_{sc} A_{sc} + 185\ A_b$$

$p_{cc} A_k + 185\ A_b$ should not exceed $0{\cdot}5\ u_w A_c$.

4. Columns with a ratio of effective length ot least lateral dimension from 15 to 57 are long columns. They can carry a maximum axial load calculated as in (b) or (c) above, reduced by the appropriate coefficient in Table 3.1.7 below (from Table 18, CP 114)

Table 3.1.7

Ratio of effective length to least lateral dimension of column	Coefficient
15	1·0
24	0·7
33	0·4
45	0·2
57	0

Linear interpolation will give intermediate values. General principles for evaluating effective column length are given in Table 19, CP 114. and Table 1.5.1, Section 1 above.

Beams and Slabs

Effective span

This is the lesser of:

1. The distance between the centres of bearings,

2. The clear distance between supports plus the effective depth of the beam.

Slender beams

If L exceeds 30b, the maximum depth in design should not exceed 8b, and the maximum compressive stress in the concrete should not exceed p_{cb} times the appropriate reduction factor in Table 3.1.8 (from Table 14, CP 114)

Table 3.1.8 Stress reduction coefficients for slender beams

Slenderness ratio L/b	30	40	50	60
Coefficient	1·0	0·75	0·5	0·25

Intermediate values may be obtained by interpolation.

Minimum reinforcement in slabs

Not less than 0·15% of gross concrete area for plain bars.
" " " 0·12% " " " " " high-yield high-bond bars.

3.2 DESIGN OF BEAMS WITH TENSILE REINFORCEMENT ONLY

Design of beams by the elastic method

Assumptions

1. Steel and concrete are elastic within the range of permissible stresses.
2. Stress is proportional to strain.

3. Variation of strain over the depth is linear.
4. The whole of the tension is taken by the reinforcement.
5. The modular ratio

$$m = \frac{\text{Young's modulus for steel}}{\text{Young's modulus for concrete}}$$

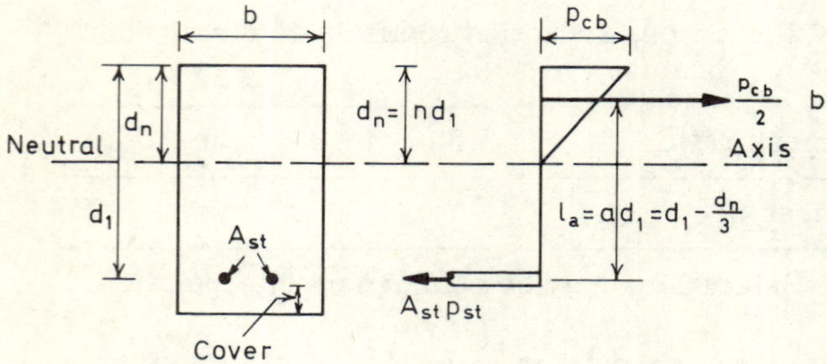

Figure 3. 2. 1.

The following expressions can be deduced, assuming a balanced section in which both concrete and steel are stressed to the permissible limits.

$$n = \frac{p_{cb}}{\left(p_{cb} + \frac{p_{st}}{m}\right)} \qquad \text{where } d_n = n\,d_1$$

$$a = 1 - \frac{n}{3} \qquad \text{where } l_a = a\,d_1$$

$$k = \frac{pcb}{2}\,.\,n.a \qquad \text{where k is a design coefficient}$$

Then Moment of resistance = Applied bending moment = $kb.d_1^2$

To design the section, calculate K, assume a suitable breadth b (not less than 1/30 span), and d_1 can be found.

Steel area can be found from

$$\text{Moment of resistance} = p_{st} A_{st} l_a$$

Hence $$A_{st} = \frac{M}{p_{st}\, a\, d_1}$$

Summary of Elastic Design

1. Calculate maximum applied bending moment (M) in N mm. The moment of resistance in concrete and steel (M_r) are designed to be equal to M.
2. $M = kbd_1^2$. Hence $d_1 = \sqrt{\left(\frac{M}{k_b}\right)}$.
3. $M = p_{st}\, A_{st}\, l_a$. Hence $A_{st} = \frac{M}{p_{st}\, a\, d_1}$, then obtain appropriate bars from Table 3.2.2.
4. Overall depth $d = d_1 + \frac{1}{2}$ diameter of bars + cover.
5. Calculate maximum shear force and design suitable shear reinforcement if necessary.
6. Check local and average bond stress and provide suitable hooks.

Elastic theory constants for rectangular beams for CP 114 nominal mixes.

Table 3.2.1.

Nominal mix	p_{cb} (N/mm²)	p_{st} (N/mm²)	n	a	k	r
1:1:2	10	140	0·517	0·828	2·14	0·0185
		125	0·546	0·818	2·24	0·0218
1:1½:3	8·5	140	0·477	0·841	1·70	0·0142
		125	0·505	0·832	1.78	0·0171
1:2:4	7	140	0·429	0·857	1·29	0·0112
		125	0·456	0·848	1·35	0·0128

$d_n = nd_1$; lever arm $= ad_1$; $M_r = kbd_1{}^2$; $A_{st} = rbd_1$.

Elastic Theory. To calculate moment of resistance (M_r) for a beam where b and d_1 are known and which contains a given amount of tensile steel (A_{st}) which may be greater or less than the economic amount.

Transformed area and moment of inertia method.

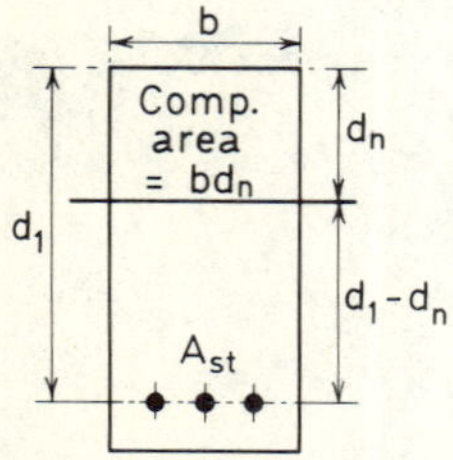

Figure 3.2.2

Table 3.2.2

Dia. (mm)	Mass (kg/m)	Areas in mm^2 for numbers of bars 1	2	3	4	5	6	7	8	9	10	11	12
6	0·222	28	57	85	113	142	170	198	226	255	283	311	340
8	0·395	50	101	151	201	252	302	352	402	453	503	553	604
10	0·616	79	157	236	314	392	471	550	628	707	785	864	942
12	0·888	113	226	339	452	566	679	792	905	1020	1130	1240	1360
16	1·579	201	402	602	804	1010	1210	1410	1610	1810	2010	2210	2410
20	2·466	314	628	943	1260	1570	1890	2200	2510	2830	3140	3560	3770
25	3·854	491	982	1470	1960	2450	2950	3440	3930	4420	4910	5400	5840
32	6·313	804	1610	2410	3220	4020	4830	5630	6430	7240	8040	8850	9650
40	9·864	1260	2510	3770	5030	6280	7540	8800	10100	11300	12600	13800	15100
50	15·413	1960	3930	5890	7850	9820	11800	13700	15700	17700	19600	21600	23600

1. Replace the steel by its transformed 'concrete area' which is m times its true area. i.e. concrete area = mA_{st}.

2. Take moments of the transformed areas about the neutral axis to determine d_n;

   i.e. $bd_n \times \frac{d_n}{2} = mA_{st}(d_1 - d_n)$

   This gives a quadratic equation which on solving gives the unknown value d_n.

3. Calculate the moment of inertia (I) about the neutral axis;

   i.e. $I = \frac{bd_n^3}{3} + mA_{st}(d_1 - d_n)^2$ (see Table 1.4.2)

4. Assuming maximum permissible concrete stress p_{cb}, calculate the moment of resistance M_r in terms of concrete;

   i.e. $M_r = \frac{p_{cb}I}{d_n}$

5. Assuming maximum permissible steel stress p_{st}, calculate the moment of resistance M_r in terms of steel;

   i.e. $M_r = \frac{p_{st}I}{m(d_1 - d_n)}$

6. The actual value of M_r is the lesser of the values from (4) and (5).

Elastic Theory. To calculate the stresses in the steel and concrete when the bending moment (M) is known, and b, d_1 and A_{st} are known.

Steps (1), (2), and (3) as in the previous paragraph.

4. Maximum concrete stress $= \frac{Md_n}{I}$

5. Maximum steel stress $= \frac{Mm(d_1 - d_n)}{I}$

Design of beams by the load factor method

Assumptions

The load factor is the ratio of the ultimate strength of a beam to its working load. This method does not assume a linear relationship between stress and strain, nor does it use the modular ratio. Tests to destruction have shown that the stress distribution in the concrete at failure is equivalent to a constant compressive stress equal to two thirds of the cube strength of the concrete acting over a depth sufficient to provide a total compression equal to the tension provided by the tensile reinforcement acting at its yield stress.

Conditions

1. Load factor is 1·8.

2. For designed mixes two thirds of the actual cube strength should be used. For nominal mixes three fifths of the actual strength should be used.

3. The stress in the steel must not exceed that given in Table 3.1.4.

4. Compressive stress in concrete two thirds of the permissible compressive stress in the concrete in bending. The depth of concrete in compression must not exceed half the effective depth.

Simplified formulae for rectangular sections without compressive reinforcement.

The moment of resistance should be the lesser of:

$$M_r = A_{st}\, p_{st}\, l_a$$

$$M_r = \frac{p_{cb}}{4}\, b\, d_1^2$$

where $l_a = d_1 - \dfrac{3\, A_{st}\, p_{st}}{4\, b\, p_{cb}}$

Load factor constants for rectangular beams for CP 114 nominal mixes

Table 3. 2. 3

Nominal mix	p_{cb} (N/mm^2)	p_{st} (N/mm^2)	k	r
1:1:2	10	140	2·5	0·0238
		125	2·5	0·0267
1:1½:3	8·5	140	2·125	0·0202
		125	2·125	0·0227
1:2:4	7	140	1·75	0·0167
		125	1·75	0·0187

$k = \dfrac{p_{cb}}{4}$

$A_{st} = \dfrac{p_{cb}}{3p_{st}} \cdot bd_1 = rbd_1$

$n = 0{\cdot}5$

$a = 0{\cdot}75$

Rectangular beams with less steel than the balanced amount by load factor method.

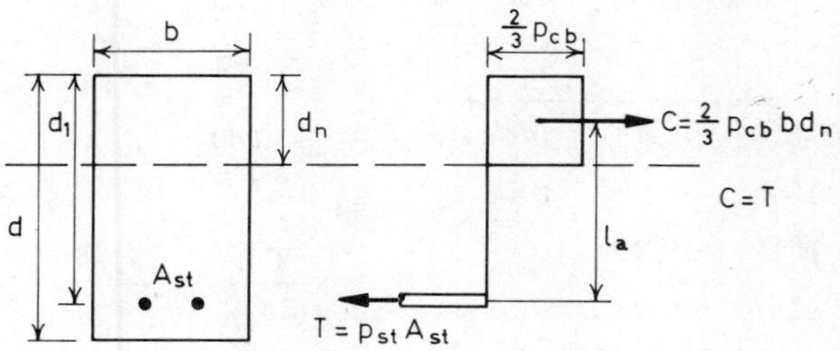

Figure 3.2.3

1. $C = T$

$\therefore \quad \frac{2}{3} p_{cb} b d_n = p_{st} A_{st}$

$\therefore \quad d_n = \frac{3}{2} \frac{p_{st} A_{st}}{p_{cb} b}$

but $A_{st} = rbd,$

$\therefore \quad n = \frac{3}{2} \frac{p_{st} r}{p_{cb}}$

2. $l_a = d_1 - \frac{d_n}{2} = d_1 - \frac{nd_1}{2}$

$\therefore \quad a = 1 - \frac{n}{2}$

3. $M_r = p_{st} A_{st} l_a = p_{st} r b d_1 a d_1 = p_{st} r a b d_1^2$

$\therefore \quad k = p_{st} s r$

Load factor constants for various values of r where $A_{st} = rbd_1$

Table 3.2.4

r	n	a	k
0·004	0·107	0·946	0·47
0·005	0·134	0·933	0·58
0·006	0·161	0·920	0·69
0·008	0·214	0·893	0·89
0·010	0·268	0·866	1·08
0·011	0·295	0·852	1·17
0·012	0·322	0·839	1·26
0·013	0·348	0·826	1·34
0·014	0·376	0·812	1·42
0·015	0·402	0·799	1·50
0·016	0·429	0·785	1·57
0·0167	0·448	0·776	1·62

$M_r = kbd_1^2$

$A_{st} = rbd_1$

$d_n = nd_1$

$$n = \frac{3p_{st}r}{2p_{cb}}$$

$l_a = ad_1$

$$a = 1 - \frac{n}{2}$$

$p_{st} = 125$

$p_{cb} = 7$

(Similar tables can be constructed for other values of p_{st} and p_{cb}.)

3.3 SLAB DESIGN

Design of slabs. Elastic theory

1. Obtain the thickness from Table 3.1.6 and calculate its mass per square metre. One square metre of reinforced concrete 10 mm thick has a mass of 24 kg.
2. Find the superimposed load in kN/m^2 from Table 1.3.1. This load plus the mass per square metre of the slab

multiplied by the gravity constant, 9·81, gives the load per metre run for design purposes for a slab one metre wide.

3. Find the maximum bending moment M(N mm) on a strip of slab one metre wide. For bending moments in continuous slabs, see Table 3.3.1.

4. Calculate the effective depth d_1 required from $d_1 = \sqrt{\left(\frac{M}{kb}\right)}$ where b = 1000 mm. k is obtained from Table 3.2.1. Thus, e.g., if $p_{st} = 125\ \text{N/mm}^2$, $p_{cb} = 7\ \text{N/mm}^2$,

$$d_1 = \sqrt{\left(\frac{M}{1{\cdot}35 \times 1000}\right)}.$$

If this value of d_1 plus an allowance for steel and cover is less than the value of d obtained from Table 3.1.6 use this latter value.

5. Find area of main steel A_{st} per metre width from

$$A_{st} = \frac{M}{p_{st} l_a}$$

and choose suitable bars and spacing from Table 3.2.2. Sizes of bars are usually from 8 mm to 20 mm diameter. Spacing should not exceed $3d_1$.

6. Calculate the area of distribution steel in mm² per metre width from

Area = 0·15% of (1000 × total depth of slab)

Choose suitable bars and spacing from Table 3.2.2. Sizes of bars are usually from 6 mm to 10 mm diameter. Spacing should not exceed $5d_1$.

Cover

Minimum cover outside main bars is 15 mm or the diameter

of the bar, whichever is greater. For external work or corrosive conditions minimum cover over all steel is 40 mm.

Shear

Shear stresses in slabs are usually small and are not usually calculated.

Design of Slabs. Load Factor Method

If the balanced values of k and l_a are used, the thickness of slab required for bending is often much less than that from Table 3.1.6. The following method is therefore suggested.

Steps (1), (2), and (3) as for elastic theory.

4. Using the value of d_1 from Table 3.1.6, calculate the value of k required from $k = M/bd_1^2$ where b = 1000 mm.
5. From Table 3.2.4, or similar tables for different stresses read off value of r corresponding to the calculated value of k.
6. Main steel per foot = rbd_1 where b = 1000 mm. Choose suitable bars.
7. Distribution steel as step (6) elastic theory.

Bending Moments in Reinforced Concrete Beams and Slabs Spanning in one Direction

Unless more exact estimates are made, the bending moments in uniformly loaded beams and slabs continuous over three or more approximately equal spans may be assumed to have the values given in Table 3.3.1.

Bending moment for design = moment due to dead load + moment due to superimposed (live) load.

Two spans may be considered as approximately equal when they do not differ by more than 15% of the longer span.

Table 3. 3. 1 is taken from CP 114 (1969).

Table 3. 3. 2 gives less accurate values based on the total load W where $W = W_d + W_s$.

Table 3. 3. 1

	Near middle of end span	At support next to end support	At middle of interior spans	At other interior supports
Moment due to dead load	$+\frac{W_d l}{12}$	$-\frac{W_d l}{10}$	$+\frac{W_d l}{24}$	$-\frac{W_d l}{12}$
Moment due to superimposed load	$+\frac{W_s l}{10}$	$-\frac{W_s l}{9}$	$+\frac{W_s l}{12}$	$-\frac{W_s l}{9}$

W_d = total dead load per span; W_s = total superimposed load per span.

Table 3. 3. 2

	Near middle of end span	At support next to end support	At middle of interior spans	At other interior supports
Moment	$+\frac{Wl}{10}$	$-\frac{Wl}{10}$	$+\frac{Wl}{12}$	$-\frac{Wl}{12}$

Slabs Spanning in Two Directions Simply Supported on Four Sides

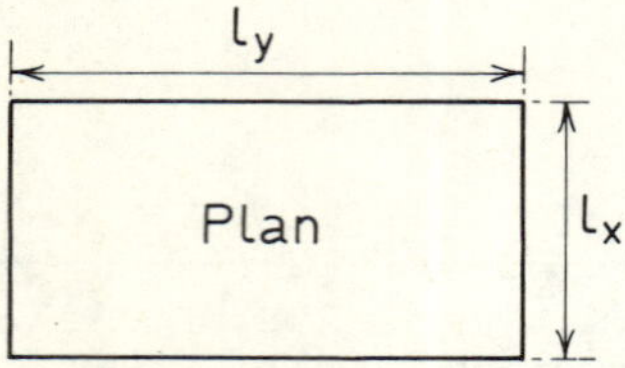

Figure 3.3.1

For the method used here the elastic theory must be used.

Design

1. Assume a thickness of slab from Table 3.1.6.

2. Estimate the total load in kN/m^2.

3. Calculate the maximum bending moments per metre width in the short and long spans from:

   Short span. $M_x = \alpha_x w l_x^2 \times 10^6$ N mm.

   Long span. $M_y = \alpha_y w l_x^2 \times 10^6$ N mm.

   w = total load on floor or roof in kN/m^2.

   l_x is the shorter span in metres.

   α_x and α_y are taken from Table 3.3.3 and these have been adjusted so that both M_x and M_y are expressed in terms of l_x.

4. Calculate the effective depth d_1 required for the short span from $d_1 = \sqrt{(M_x/kb)}$ where b = 1000 mm and k = balanced value (Table 3.2.1). Adjust d_1 if necessary as step (4) of elastic theory.

5. Calculate the effective depth d_1 required for the long span

from $d_1 = \sqrt{(M_y/kb)}$ where b = 1000 mm. Often this calculated value of d_1 is much less than the actual value resulting by placing the steel on top of the short span steel. In this case the actual value of d_1 for the calculation of the long span steel will be d_1 for the short span steel less the bar diameter.

6. Calculate the short span steel per metre width from $A_{st} = \dfrac{M_x}{p_{st} l_a}$ where $l_a = a_1 \times$ (d_1 for short span). Choose suitable bars and spacing from Table 3. 2. 2.

7. Calculate the long span steel per metre width from
$$A_{st} = \frac{M_y}{p_{st} l_a}$$
where $l_a = a_1 \times$ (d_1 for long span) and choose suitable bars and spacing.

8. When the slab is square, $\alpha_x = \alpha_y = 0{\cdot}062$, therefore $M_x = M_y$. The calculated depth d_1 must therefore be provided for the upper layer of steel and the overall depth of slab = d_1 + 1½ times bar diameter + cover.

Table 3. 3. 3. Bending Moment Coefficients for Slabs spanning in two directions simply supported on four sides

l_y/l_x	1·0	1·1	1·2	1·3	1·4
α_x	0·062	0·074	0·084	0·093	0·099
α_y	0·062	0·061	0·059	0·055	0·051
l_y/l_x	1·5	1·75	2	2·5	3
α_x	0·104	0·113	0·118	0·122	0·124
α_y	0·046	0·037	0·029	0·020	0·014

When tables are not available, proceed as follows:

Divide the total load w kN/m^2 between the two directions in the ratio of $\left(\frac{l_y}{l_x}\right)$.

For example,

$$l_y = 6 \text{ m} \quad l_x = 3 \text{ m}$$

$$\left(\frac{l_y}{l_x}\right)^4 = \left(\frac{6}{3}\right)^4 = \left(\frac{2}{1}\right)^4 = \frac{16}{1}$$

$$M_x = \frac{16}{17}\frac{wl^2{}_x}{8} \times 10^6 \text{ Nmm}$$

$$M_y = \frac{1}{17}\frac{wl^2{}_y}{8} \times 10^6 \text{ Nmm}$$

w = total load in kN/m^2

l_x and l_y are in metres

Note that in this method the value of l_x is used only for calculating M_x. l_y is used when calculating M_y.

Continuous Slabs Spanning in Two Directions

The bending moments are less than in simply supported slabs. This case is too complicated to be dealt with here. CP114 gives a whole page of bending moment coefficients for use in different conditions.

3.4 DESIGN OF RECTANGULAR BEAMS WITH TENSILE AND COMPRESSIVE STEEL

Elastic theory

To determine the areas of steel A_{st} and A_{sc} to resist a given bending moment M when the dimensions b, d_1 and d_2 are known.

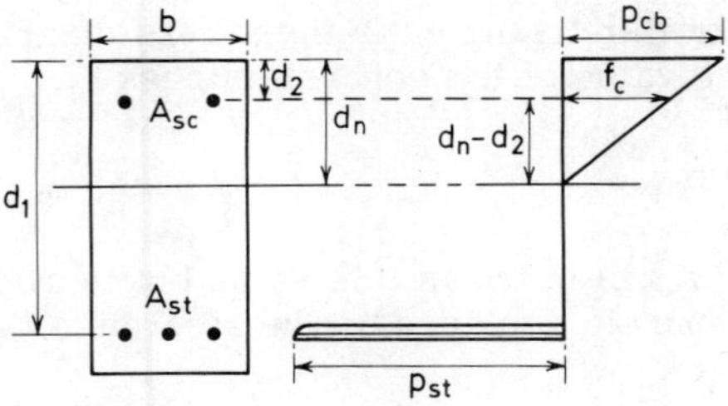

Figure 3.4.1

1. Assume the economic section and maximum stresses p_{cb} and p_{st}. Calculate the economic moment of resistance M_e and the corresponding values of d_n and l_a (Table 3.2.1.) e.g., when $p_{cb} = 7\ N/mm^2$; $p_{st} = 125\ N/mm^2$ m = 15; Me = $1{\cdot}35\ bd_1^2$; $d_n = 0{\cdot}456 d_1$; $l_a = 0{\cdot}848 d_1$.

2. Calculate the area of tensile steel from:

$$A_{st} = \frac{M_e}{p_{st} l_a} + \frac{M - M_e}{p_{st}(d_1 - d_2)}$$

e.g., for 7-125-15 concrete

$$A_{st} = \frac{1{\cdot}35 bd_1^2}{125 \times 0{\cdot}848 d_1} + \frac{M - 1{\cdot}35 bd_1^2}{125(d_1 - d_2)}$$

Alternatively,

$$A_{st} = \frac{M}{125 \times 0{\cdot}848 d_1}$$

Both methods are slightly approximate.

3. Calculate from similar triangles in the stress diagram (Figure 3.4.1) the value of the concrete stress f_c,

   $$\text{i.e. } f_c = \frac{d_n - d_2}{d_n} p_{cb}$$

   The steel stress (f_s) is m times this value but to allow for the area of concrete replaced by the steel take $f_s = (m - 1)f_c$).

4. Calculate the area of compressive steel from

   $$A_{sc} = \frac{M - M_e}{f_s(d_1 - d_2)} = \frac{M - M_e}{(m - 1)f_c(d_1 - d_2)}$$

Elastic Theory

To calculate the moment of resistance of a given rectangular cross-section beam containing known amounts of tensile and compressive steel.

Transformed area and moment of inertia method.

'Concrete area' of $A_{st} = mA_{st}$.

Effective 'concrete area' of $A_{sc} = (m - 1)A_{sc}$.

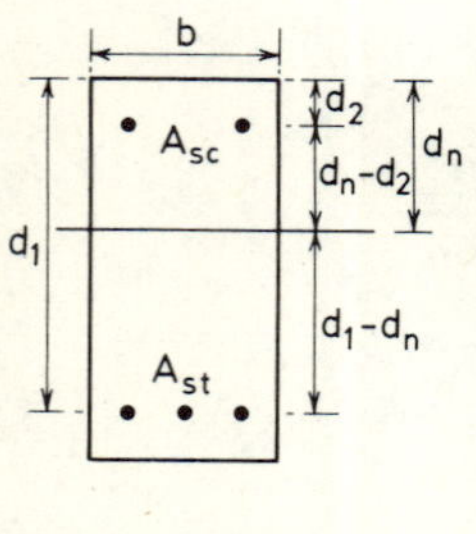

Figure 3.4.2

1. Take moments of transformed areas about neutral axis to determine d_n i.e. $bd_n \times \frac{d_n}{2} + (m-1)A_{sc}(d_n - d_2) = mA_{st}(d_1 - d_n)$.

   This gives a quadratic equation which on solving gives d_n.

2. Calculate the moment of inertia (I) in concrete units about the neutral axis.

   $$\text{i.e. } I = \frac{bd_n^3}{3} + (m-1)A_{sc}(d_n - d_2)^2 + mA_{st}(d_1 - d_n)^2$$

3. Assuming maximum permissible concrete stress p_{cb} calculate the moment of resistance from

   $$M_r = \frac{p_{cb}I}{d_n}$$

4. Assuming maximum permissible steel stress p_{st} calculate the moment of resistance from

   $$Mr = \frac{P_{st}I}{m(d_i - d_n)}$$

5. The actual value of M_r is the lesser of the values from (3) and (4).

Elastic Theory

To calculate the stress in the steel and concrete when the bending moment (M) is known and b; d_1; d_2; A_{st} and A_{sc} are known.

Steps (1) and (2) as above.

3. Stress in tensile steel $= \frac{Mm(d_1 - d_n)}{I}$.

4. Maximum stress in concrete $= \frac{Md_n}{I}$.

5. Stress in compressive steel = maximum stress in concrete multiplied by $\dfrac{m(d_n - d_2)}{d_n}$.

Load Factor Method
Rectangular beams with tensile and compressive steel.

For doubly reinforced beams the load factor method is more economical than the elastic theory. Recall that the maximum moment of resistance of a beam with tensile steel only is, by the load factor method,

$$M_r = \frac{p_{cb}bd_1^2}{4}.$$

If a greater moment of resistance is required, the compressive resistance of the beam must be increased by putting steel in the compressive zone. When compressive steel is required the size of the beam will be known and the required calculations are to determine the amounts of steel A_{st} and A_{sc}.

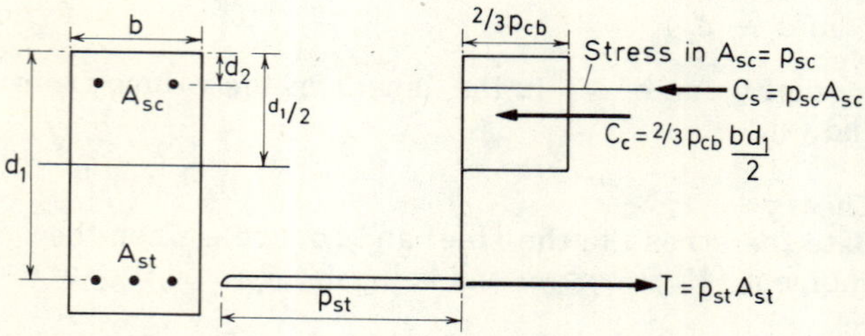

p_{sc} = permissible compressive stress in steel (see Table 3.1.4)

Figure 3.4.3

C_s = compression taken by the concrete; C_c = compression taken by the steel.

Note that there is no need to allow for the small reduction in concrete area by the amount displaced by the compressive reinforcement. (As stated in CP 114.)

Lever arm for $C_s = d_1 - d_2$. Lever arm for $C_c = \frac{3}{4}d_1$.

1. Calculate the maximum bending moment M(Nmm)

2. $$M = \frac{p_{cb}bd_1^2}{4} + A_{sc}p_{sc}(d_1 - d_2)$$

   This equation gives the area A_{sc}.

3. Calculate the total compression C from

   $$C = \frac{2p_{cb}}{3}\frac{bd_1}{2} + A_{sc}p_{sc}$$

   This equals the total tension, T.

4. Calculate the area of tensile steel from

   $$A_{st} = \frac{T}{p_{st}}$$

Note (1). The compression reinforcement should be effectively anchored in two directions at right angles at points not further apart than twelve times the diameter of the anchored bar. This subsidiary reinforcement should pass round or be hooked over both the compressive and tensile reinforcement.

Note (2). A_{sc} should preferably not exceed 4% of the total cross-sectional area of the beam. If it does, only 4% should be allowed for in calculations.

3.5 DESIGN OF T-BEAMS AND L-BEAMS

Load factor method

This type of beam usually occurs when the beam is monolithic with the floor slab. Since there is usually a large amount of concrete available to take compression the beam could be of small depth, but for practical purposes (deflection etc.) it is usual to assume an overall depth and breadth of rib as follows.

Overall Depth. About $\frac{1}{15}$ to $\frac{1}{20}$ span for light loads such as roofs.

About $\frac{1}{15}$ span for moderate floor loads.

About $\frac{1}{12}$ span for heavy floor loads.

Width of rib = (b_r) = $\frac{1}{3}$ to $\frac{1}{2}$ of overall depth of beam.

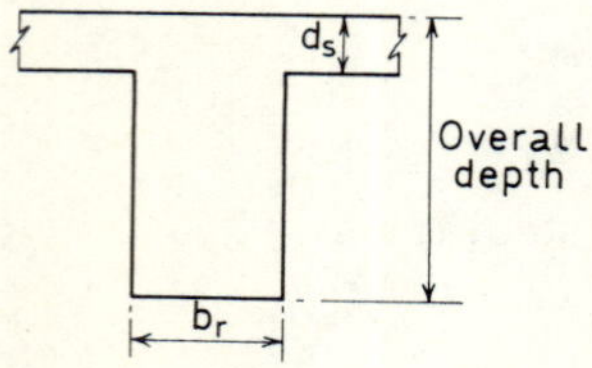

Figure 3.5.1

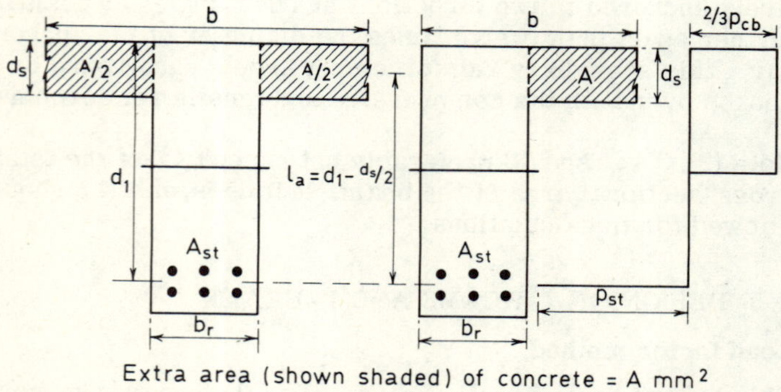

Figure 3.5.2

Design

1. Assume an overall depth and breadth of rib and estimate weight of beam for adding to design loads.

2. Calculate the maximum bending moment M (N mm)

3. Deduct about 60 or 75 mm from the overall depth to allow for cover and two layers of bars $\therefore$ d_1 = overall depth minus 60 or 75 mm.

4. For T- and L- beams, the lever arm l_a can be taken as $d_1 - d_s/2$ where d_s is the thickness of the flange (i.e. floor slab).

5. Calculate A_{st} from

$$A_{st} = \frac{M}{p_{st} l_a} = \frac{M}{p_{st}\left(d_1 - \frac{d_s}{2}\right)}$$

6. Determine permissible value of b and estimate area A. This permissible value of b is the least of the values given in Table 3.5.1.

Table 3.5.1

T-beams	L-beams
(a) $\frac{1}{3}$ effective span of beam	$\frac{1}{6}$ effective span of beam
(b) Distance between centres of ribs of beams	Breadth of rib + $\frac{1}{2}$ of clear distance between ribs
(c) Breadth of rib + 12 times thickness of slab	Breadth of rib + 4 times thickness of slab

7. Check whether compressive M_r of beam is adequate from

$$M_r = \frac{p_{cb} b_r d_1^2}{4} + \frac{2}{3} p_{cb} A l_a = \frac{p_{cb} b_r d_1^2}{4} + \frac{2}{3} p_{cb} d_s (B - b_r)\left(d_1 - \frac{d_s}{2}\right)$$

If this value is equal to or exceeds the bending moment M the beam is adequate in compression. If it is less, the beam must be made larger or compression steel supplied.

8. Calculate shear stresses and design shear reinforcement.

Note. If the thickness of slab (d_s) exceeds $d_1/2$ the value of $d_1/2$ must be used instead of d_s in the equation of step 7.

Elastic Theory

To calculate the moment of resistance of a given T-beam with a known amount of tensile steel.

Transformed area and moment of inertia method.

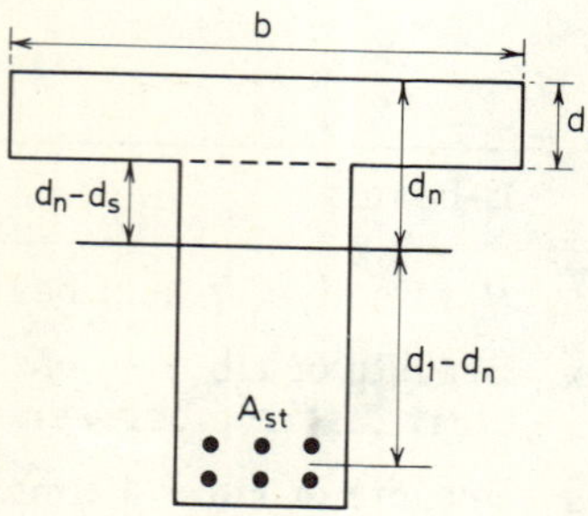

Figure 3.5.3

1. Take moments of transformed areas about the neutral axis

$$\text{i.e. } bd_s\left(d_n - \frac{d_s}{2}\right) + b_r(d_n - d_s)\left(\frac{d_n - d_s}{2}\right)$$

$$= mA_{st}(d_1 - d_n)$$

This is a quadratic equation which on solving gives d_n. (If d_n comes out less than d_s which means that the neutral axis is within the flange the beam can be taken as a rectangular beam of breadth b.)

2. Calculate the moment of inertia I about the neutral axis. i.e. if d_n from step 1 is greater than d_s,

$$I = \left[\frac{bd_s^3}{12} + bd_s\left(d_n - \frac{d_s}{2}\right)^2\right] + b_r\,\frac{(d_n - d_s)^3}{3}$$

$$+ mA_{st}(d_1 - d_n)^2$$

3. Assuming maximum permissible concrete stress p_{cb} calculate the moment of resistance M_r in terms of concrete.

$$\text{i.e. } M_r = \frac{p_{cb}I}{d_n}$$

4. Assuming maximum permissible steel stress p_{st} calculate M_r in terms of steel.

$$\text{i.e. } M_r = \frac{p_{st}I}{m(d_1 - d_n)}$$

5. The actual value of M_r is the lesser of the values from (3) and (4).

3.6 DESIGN OF HOLLOW TILE FLOORS

Load factor method

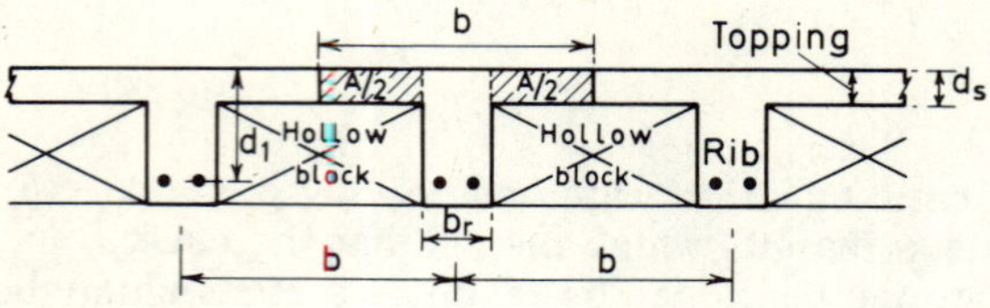

Figure 3.6.1

Usually the permissible breadth b of the flange is equal to the distance between centre of ribs.

Thickness of topping usually 40 to 50 mm.

Width of rib usually 75 to 100 mm.

In the treatment here given the hollow blocks are not regarded as contributing to the structural strength.

Design

1. Use Table 3.1.6 to obtain minimum depth of floor and assume a thickness of topping (d_s) and breadth of rib (b_r).
2. Estimate the total load on the span for a breadth b of floor. (This load should include the weight of the T-beam.) Calculate the maximum bending moment M (N mm).
3. Determine d_1 i.e. overall depth minus (cover plus half bar diameter) and l_a,

   i.e. $d_1 - \dfrac{d_s}{2}$

4. Calculate area of steel for one rib from

$$A_{st} = \frac{M}{p_{st} l_a}$$

5. Calculate compressive M_r as step (7) for T-beams. If this value is less than the bending moment M or if A_{st} is too large for accommodation in the rib, assume a deeper floor and repeat the calculations.

Shear and Bond

Having calculated the beam for bending, the shear stress (q) at any cross-section is calculated from $q = Q/bl_a$. Q = the shear force, b = breadth of rectangular beam. For T and L beams b_r should be used instead of b.

If the calculated value of q does not exceed the permissible value (from Table 3.1.1) shear reinforcement is not required but stirrups at nominal spacing may be provided. When the calculated value of q exceeds the permissible value, vertical stirrups or inclined bars or both may be used to take the whole of the shear force. Even when shear reinforcement is provided the value of q from equation above should not exceed four times the permissible shear stress for the concrete alone. The design of members without shear reinforcement should be limited to slabs, footings and members of secondary importance.

CP 114: 1969 states 'The design of members without shear reinforcement should be limited to slabs, footings, and members of secondary importance. Even for such members a conservative approach should be used in calculating the shear resistance of the members when no shear reinforcement is provided since recent research has indicated that, in some circumstances, the margin of safety may be lower than desirable.'

This means that for all beams (with the possible exception of small lintels) shear reinforcement must be provided even if the calculated shear stress is less than the permissible value for plain concrete. CP 114 does not give any guidance as to the amount of this shear reinforcement but some authorities are assuming that the 'nominal' reinforcement should be equal in area to not less than 0·15 per cent of the horizontal area of the concrete, e.g. area of nominal shear reinforcement in square millimetres per metre length of beam = $0{\cdot}0015 \times 1000 \times b$ where b is the breadth of the beam (or breadth of rib in the case of T-beam or L-beam).

Neither the longitudinal spacing of the nominal shear reinforcement nor the lateral spacing of the vertical legs of the stirrups should exceed a distance equal to the lever arm.

3.7 SIMPLE COLUMN BASES

The thickness of the base must be sufficient to resist safely the shear forces and bending moments. The 'punching shear' method of calculating the thickness has been much used but is not mentioned in CP 114. If this method is used, the permissible punching shear stress can be taken as twice the beam shear stress given in Table 3.1.1.

Design

Area of base = $A = L_1 L_2$ (mm^2).

Area of column = a (mm^2).

Shaded area = $A - a$ (mm^2) (Figure 3.7.1).

Column load = W (newtons).

Weight of base = w (newtons).

Upthrust of soil per mm^2 of base = $\frac{W}{A}$ N (ignoring weight of base).

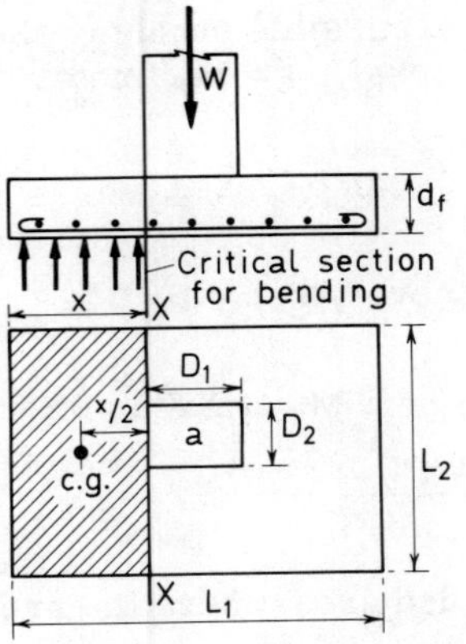

Figure 3.7.1

1. Determine approximate weight of base w and calculate area of base from

$$A = \frac{W + w}{p_b} \quad \text{(A in mm}^2\text{; W and w in newtons; } p_b \text{ in N/mm}^2\text{)}$$

where p_b = permissible bearing pressure on soil. Calculate L_1 and L_2 (if base is square $L_1 = L_2$).

2. Using punching shear method,

$$d_f = \frac{W}{p_r q_p}\left(\frac{A - a}{A}\right)$$

d_f = total depth of base. q_p = permissible punching shear stress = (2 × ordinary shear stress), p_r = perimeter of column.

As a very close approximation d_f can be taken as $W/p_r q_p$.

Deduct, say 60 mm to allow for cover, then effective depth, $d_1 = d_f$ − about 60 mm.

3. Calculate maximum bending moment M_x at XX as follows.

Upthrust per mm² of base $= \dfrac{W(N)}{A\ (mm^2)}$.

Load on shaded area $= \dfrac{W}{A} \times$ shaded area, where the area is in mm².

M_x = Load on shaded area $\times \dfrac{x}{2}$ where x is in millimetres.

$$\therefore \quad M_x(N\ mm) = \frac{W}{A} \times \text{shaded area} \times \frac{x}{2}$$

W = column load (N) A and shaded area are in mm² x is in mm ($A = L_1 L_2$).

4. Calculate the effective depth required to resist the bending moment from $d_1 = \sqrt{(M/KL_2)}$ where K is the factor from Table 3.2.1.

5. For calculating the steel at right angles to the length L_2 use the greater value of d_1 from (2) and (4).

6. A_{st} at right angles to $L_2 = \dfrac{M_x}{p_{st} l_a}$.

7. Calculate the bending moment about YY (Figure 3.7.2) from

$$M_y = \frac{W}{A} \times \text{shaded area} \times \frac{y}{2}$$

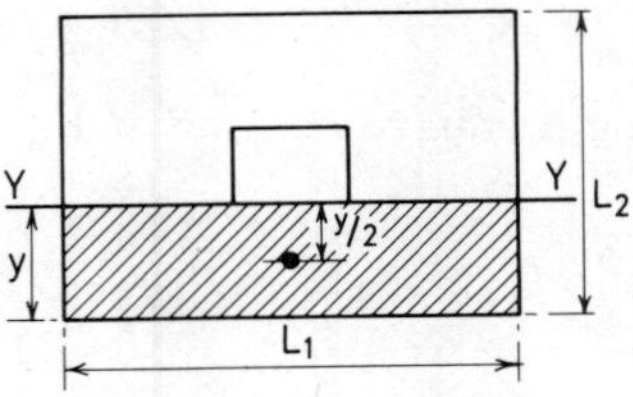

Figure 3.7.2

Calculate the steel from

$$A_{st} = \frac{M_y}{p_{st} l_a}$$

If column is square and foundation is square, then $M_x = M_y$ and the area of steel in each direction is the value from paragraph (6). In rectangular bases for columns, the reinforcement parallel to the short edge should be more closely spaced near the column.

(8) There is usually no need to calculate the 'beam shear' stress when the punching shear method of design is used, but bond stresses may be calculated, and small diameter bars used to prevent the bond stress being excessive.

Considering axis XX, Figure 3.7.1, load (i.e. shear) for calculating the local bond stress equals the upthrust on the shaded area.

Since local bond stress $= \dfrac{Q}{l_a o}$ it is $\dfrac{\text{upthrust on shaded area}}{l_a o}$

where o is the sum of the circumferences of the bars in the width L_2.

Design, ignoring Punching Shear

Example of calculations assuming a square column and a square base.

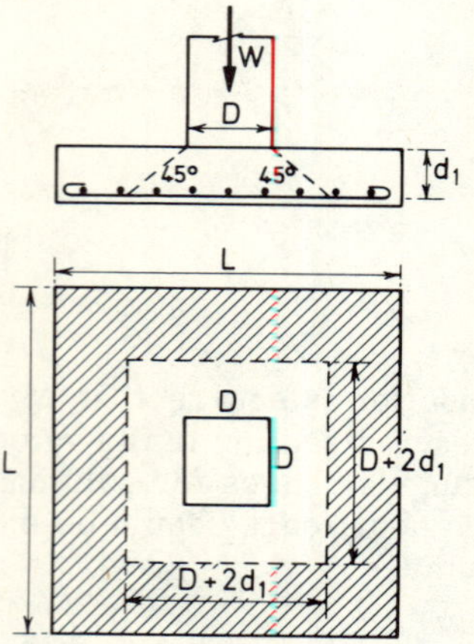

Figure 3.7.3

9. Calculate area of base and dimension L as step (1).
10. Calculate maximum bending moment as step (3).
11. Calculate d_1 as step 4.
12. Check for shear.

Load causing shear = Upthrust on shaded area

$$= \frac{W}{L^2}[L^2 - (D + 2d_1)^2 \qquad \text{(a)}$$

breadth of 'beam' resisting shear = $4(D + 2d_1)$.

$$q = \frac{Q}{bl_a}$$ where Q = value (a) from step 12, and b = the

perimeter, i.e. $4(D + 2d_1)$.

If the calculated value of q exceeds the permissible value assume a greater value of d_1 and repeat the shear calculations.

13. Calculate the area of steel as step (6).

14. Calculate the local bond stress as step (8).

Section 4
Structural Timber

Design Requirements
Design of Beams and Struts
Timber Sizes for Small Buildings

4. Structural Timber

4.1 DESIGN REQUIREMENTS FROM CP 112, 'THE STRUCTURAL USE OF TIMBER.'

Many of the requirements of CP 112 are too complicated to be summarised in a book of this description and apart from design of simple beams and struts reference must be made to the code itself.

Grade stresses

Permissible stresses in timber are related to its grade. The placing of a piece of timber in the 75, 65, 50, or 40 grade depends upon its freedom from visible defects such as knots, fissures, slope of grain etc., which are specified in the code. The appropriate stresses are then related to a basic stress for a timber free from all defects.

Species groups

Stresses for individual timbers are given, but Table 4.1.1 gives softwood species in groups with similar properties and Table 4.1.2 gives the appropriate stresses.

Table 4.1.1 Softwood species groups

Species group	Name of timber	
S1	Douglas fir Pitch pine	Larch
S2	Western hemlock Parana pine Redwood	Whitewood Canadian spruce Scots pine
S3	European spruce Sitka spruce	Western red cedar

Table 4.1.2 Green stresses and moduli of elasticity for grouped softwoods in N/mm^2

Species group	Grade	Bending and tension parallel to grain	Compression parallel to grain	Compression perpendicular to grain	Shear parallel to grain	Modulus of elasticity Mean	Modulus of elasticity Minimum
S1	Basic	13·8	9·7	1·72	1·38		
	75	10·3	7·2	1·52	1·03		
	65	9·0	6·2	1·52	0·90	9000	4500
	50	6·9	4·8	1·31	0·69		
	40	5·5	3·8	1·31	0·55		
S2	Basic	11·0	8·3	1·38	1·38		
	75	8·3	6·2	1·17	1·03		
	65	6·9	5·2	1·17	0·90	6900	4100
	50	5·5	4·1	1·03	0·69		
	40	4·5	3·1	1·03	0·55		
S3	Basic	7·6	5·5	1·03	1·10		
	75	5·5	4·1	0·90	0·83		
	65	4·8	3·4	0·90	0·69	5900	3100
	50	3·8	2·7	0·76	0·55		
	40	3·1	2·07	0·76	0·41		

Note These stresses apply to timber with a moisture content greater than 18%.

Modification factors

Duration of loading

For medium and short term loading permissible stresses can be increased by the factors from Table 4.1.3.

Table 4.1.3 Modification of stresses in flexural members for duration of loading

Duration of loading	Factor
Long term (e.g. dead + permanent imposed)	1·00
Medium term (e.g. dead + snow, dead + temporary loads)	1·25
Short term (e.g. dead + imposed + snow + wind)	1·50

Slenderness of compression members

The permissible stresses in posts and struts depend upon the slenderness ratio l/r, where l is the effective length (see Table 1.5.1) and r the radius of gyration. For rectangular members the ratio l/b may be used, where b is the least lateral dimension. Table 4.1.4 gives modification factors for grades 40 and 50 softwoods and Tables 4.1.5 and 4.1.6 give the appropriate stresses for these timbers from groups S1 and S2. CP 112 gives formulae for calculating stresses for grades 65 and 75 softwoods and hardwoods.

The maximum slenderness ratio (l/r) should be 180 for a member carrying dead and superimposed loads and 250 for a member carrying only wind loads.

Table 4.1.4 Modification factors for stresses in compression members of grade 40 or grade 50 softwood

Slenderness ratio			Modification factors		
	l/r	*l*/b	Long-term loads	Medium-term loads	Short-term loads
Less	5	1·4	1·00	1·25	1·50
than	5	1·4	0·99	1·24	1·49
	10	2·9	0·98	1·23	1·47
	20	5·8	0·96	1·20	1·44
	30	8·7	0·94	1·17	1·40
	40	11·5	0·91	1·13	1·34
	50	14·4	0·87	1·08	1·27
	60	17·3	0·83	1·00	1·16
	70	20·2	0·77	0·90	1·01
	80	23·0	0·70	0·79	0·86
	90	26·0	0·61	0·68	0·72
	100	28·8	0·53	0·58	0·60
	120	34·6	0·40	0·42	0·44
	140	40·4	0·31	0·32	0·33
	160	46·2	0·24	0·25	0·25
	180	52·0	0·20	0·20	0·20
	200	57·7	0·16	0·16	0·17
	220	63·5	0·13	0·14	0·14
	240	69·2	0·11	0·12	0·12
	250	72·2	0·10	0·11	0·11

Table 4.1.5 Permissible compressive stresses parallel to the grain in N/mm^2 for 40 grade and 50 grade softwood from group S1

Slenderness ratio		Permissible compressive stress parallel to grain					
		Long term loads		Medium term loads		Short term loads	
l/r	*l*/b	40 grade	50 grade	40 grade	50 grade	40 grade	50 grade
5	1·4	3·80	4·80	4·75	6·00	5·70	7·20
10	2·9	3·72	4·70	4·66	5·90	5·59	7·05
20	5·8	3·64	4·60	4·55	5·76	5·46	6·90
30	8·7	3·56	4·50	4·44	5·60	5·32	6·71
40	11·5	3·44	4·35	4·30	5·42	5·09	6·42
50	14·4	3·30	4·17	4·10	5·18	4·82	6·10
60	17·3	3·15	3·96	3·80	4·80	4·40	5·55
70	20·2	2·92	3·70	3·42	4·32	3·84	4·85
80	23·0	2·66	3·36	3·00	3·79	3·76	4·14
90	26·0	2·32	2·92	2·58	3·26	2·73	3·46
100	28·8	2·00	2·55	2·20	2·78	2·27	2·88
120	34·6	1·52	1·92	1·60	2·02	1·67	2·11
140	40·4	1·16	1·48	1·22	1·54	1·24	1·58
160	46·2	0·90	1·15	0·95	1·20	0·95	1·20
180	52·0	0·76	0·96	0·76	0·96	0·76	0·96
200	57·7	0·61	0·77	0·61	0·77	0·65	0·82
250	72·2	0·38	0·48	0·42	0·53	0·42	0·53

Note: The values in this table are based on stresses of 3·8 N/mm^2 for 40 grade and 4·8 N/mm^2 for 50 grade timber modified by the appropriate factors from Table 4.1.4.

Table 4.1.6 Permissible compressive stresses parallel to the grain in N/mm^2 for 40 grade and 50 grade softwood from group S2.

Slenderness ratio		Permissible compressive stress parallel to grain					
		Long term loads		Medium term loads		Short term loads	
l/r	*l*/b	40 grade	50 grade	40 grade	50 grade	40 grade	50 grade
5	1·4	3·10	4·10	3·88	5·12	4·65	6·15
10	2·9	3·04	4·02	3·80	5·04	4·55	6·02
20	5·8	2·97	3·94	3·72	4·92	4·46	5·90
30	8·7	2·91	3·86	3·63	4·80	4·34	5·74
40	11·5	2·82	3·72	3·50	4·62	4·15	5·50
50	14·4	2·69	3·56	3·35	4·43	3·94	5·20
60	17·3	2·56	3·40	3·10	4·10	3·59	4·75
70	20·2	2·38	3·16	2·79	3·68	3·13	4·15
80	23·0	2·17	2·87	2·45	3·24	2·67	3·52
90	26·0	1·89	2·50	2·12	2·78	2·23	2·95
100	28·8	1·66	2·17	1·80	2·37	1·86	2·46
120	34·6	1·24	1·64	1·30	1·72	1·36	1·80
140	40·4	0·96	1·27	0·99	1·31	1·02	1·35
160	46·2	0·74	0·98	0·78	1·02	0·78	1·02
180	52·0	0·62	0·82	0·62	0·82	0·62	0·82
200	57·7	0·50	0·66	0·50	0·66	0·53	0·70
250	72·2	0·31	0·41	0·34	0·45	0·34	0·45

Note: The values in this table are based on stresses of 3·10 N/mm^2 for 40 grade and 4·10 N/mm^2 for 50 grade timber modified by the appropriate factors from Table 4.1.4.

Effective span

The span of flexural members is the distance between the centres of bearings.

Deflection

Deflection should not exceed 0·003 of the span. Table 4.1.7 gives values of span to depth ratios which satisfy this requirement.

Table 4.1.7

Softwood group	Grade	f_{ppar} N/mm^2	E (mean) N/mm^2	$\frac{\text{Span}}{\text{Depth}}$
S1	50	6·9	9000	18·8
	40	5·5	9000	23·5
S2	50	5·5	6900	17·8
	40	4·5	6900	21·8

Load sharing

If four or more rafters, joists or trusses are spaced at 600 mm or closer, and the load is distributed by purlins, boarding, etc. grade stresses should be multiplied by 1·1.

Lateral support

Table 4.1.8 gives maximum depth to breadth ratios for various degrees of lateral support.

Table 4.1.8

Degree of lateral support	Maximum depth to breadth ratio
None	2
Ends held	3
Ends held and member held in line by purlins etc.	4
Ends held and compression edge held in line by joists etc.	5
Ends held, compression edge held in line by joists etc. and bridging at spacing not greater than 6 depths	6
Ends and both edges firmly held.	7

Compression members subject to bending

In such a member $\frac{f_{apar}}{f_{ppar}} + \frac{c_{apar}}{c_{ppar}}$

should not exceed 1·0 if the slenderness ratio (l/r) is less than 20, and should not exceed 0·9 if the slenderness ratio exceeds 20.

4.2 DESIGN OF BEAMS AND STRUTS

Beams of solid rectangular cross-section

1. Calculate the maximum bending moment M in N mm

2. $M = f_{ppar} Z$

   where f_{ppar} is the permitted bending stress from Table 4.1.2.

   $$Z = \frac{bd^2}{6}$$

   $$\therefore M = f_{ppar} \frac{bd^2}{6}$$

   hence $$d^2 = \frac{6M}{bf_{ppar}}$$

   $$\therefore d = \sqrt{\left(\frac{6M}{bf_{ppar}}\right)}$$

3. A suitable dimension is chosen for b from the range of standard sections and d is found. The nearest standard section with a depth greater than the design depth is chosen from Table 4.2.1.

4. The dimensions of the beam are then checked for deflection (Table 4.1.7 or calculation) and for depth to breadth ratio depending on lateral support (Table 4.1.8).

Strut of rectangular section carrying an axial load

1. Calculate the load W in newtons.
2. Choose a trial section.
3. Calculate the slenderness ratio (l/b) and find the appropriate stress from Tables 4.1.5 or 4.1.6. (These tables cover most normally used timbers but otherwise refer to the code.)
4. Multiply the cross-sectional area and the permitted stress and compare with the load W. By a process of trial and error a suitable section is obtained.

Table 4.2.1 Sawn softwood sizes

mm	75	100	125	150	175	200	225	250	300
16	x	x	x	x					
19	x	x	x	x					
22	x	x	x	x					
25	x	x	x	x	x	x	x	x	x
32	x	x	x	x	x	x	x	x	x
38	x	x	x	x	x	x	x		
44	x	x	x	x	x	x	x	x	x
50	x	x	x	x	x	x	x	x	x
63		x	x	x	x	x	x		
75		x	x	x	x	x	x	x	x
100		x		x		x		x	x
150				x		x		x	x
200						x			
250								x	
300									x

The sizes within the lines are normally but not exclusively of European origin. The larger sizes outside the lines are normally but not exclusively of American origin.

4.3 TIMBER SIZES FOR SMALL BUILDINGS

Tables 4.3.2 to 4.3.7 give suitable sizes for timber members in domestic type buildings, based on the information in Table 4.3.1.

Table 4.3.1

Member	Dead load kg/m²	Imposed load kN/m²
Floor joists	—	1·5
floor boards 25 mm	12·25	—
9·5 mm gypsum plaster-board (solid core)	8·3	—
Ceiling joists	—	0·75
ceiling including thermal insulation	19·4	—
Rafters and purlins for pitched roofs covered with tiles or slates	—	see CP3 Chapter V
clay plain tiles (102 mm gauge)	63·5	—
concrete plain tiles (102 mm gauge)	68·4	—
Welsh slating, 75 mm lap	48·8	—
interlocking tiles, single lap	39·1 ± 4·9 kg/m²	—
felt under tiles or slates	3·6	—
thermal insulation, 25 mm thick	36·6	—
Purlins supporting sheeting for roofs having a pitch of 22½° or more	—	see CP3 Chapter V
asbestos cement corrugated sheeting	16·6	—

Table 4.3.2

Minimum sizes for various flooring and roofing members

	mm
Wall plates	50 × 75
Herringbone strutting	25 × 50
Solid bridging	22, 25, & 50 thick
Ridges	25 × 125
Hips	25 × 175
Valleys	32 thick
Fascia boards	19 × 125

Slating and tiling battens... plain tiling:
20 × 25 mm battens for rafters up to 460 mm centres
30 × 25 mm battens for rafters over 460 and up to 610 mm centres

Slating and interlocking (single lap) tiling:
20 × 40 mm battens for rafters up to 460 mm centres
25 × 40 mm up to 610 mm centres

Snow boarding for valley gutter:
four lines of 25 × 50 mm or 38 × 50 mm battens on 25 × 75 mm bearers up to 760 mm centres

Tables 4.3.3 Floor joists

Dead load in kg/m^2 supported by the joists excluding the weight of the joists

Basic size of joist (mm)	Note more than 50				51 to 75 incl.				76 to 100 incl.				101 to 125 incl.			
	Spacing of joists in mm for centres up to															
	400	450	525	600	400	450	525	600	400	450	525	600	400	450	525	600
	Maximum span of joist in metres															
38 × 75	0·940	0·838	0·737	0·635	0·889	0·813	0·686	0·610	0·838	0·760	0·660	0·584	0·813	0·737	0·635	0·533
100	1·630	1·450	1·270	1·120	1·500	1·350	1·190	1·070	1·400	1·270	1·120	1·020	1·320	1·220	1·070	0·965
125	2·410	2·180	1·900	1·700	2·180	1·980	1·750	1·570	2·018	1·830	1·630	1·470	1·880	1·730	1·520	1·400
150	2·840	2·690	2·490	2·260	2·670	2·510	2·290	2·060	2·510	2·360	2·110	1·900	2·390	2·180	1·960	1·780
50 × 75	1·190	1·070	0·940	0·813	1·120	1·020	0·889	0·787	1·070	0·965	0·838	0·762	1·020	0·914	0·813	0·737
100	2·010	1·830	1·600	1·420	1·850	1·700	1·500	1·320	1·730	1·570	1·400	1·240	1·630	1·470	1·320	1·190
125	2·640	2·540	2·390	2·130	2·540	2·440	2·160	1·960	2·440	2·240	2·010	1·800	2·260	2·080	1·880	1·700
150	3·120	3·000	2·840	2·670	3·000	2·870	2·670	2·490	2·870	2·720	2·510	2·310	2·720	2·570	2·360	2·160
175	3·660	3·510	3·330	3·120	3·510	3·350	3·120	2·920	3·350	3·180	2·950	2·770	3·200	3·020	2·790	2·620
200	4·170	4·040	3·810	3·560	4·010	3·840	3·560	3·350	3·840	3·630	3·380	3·150	3·660	3·450	3·200	3·000
63 × 150	3·350	3·230	3·070	2·950	3·230	3·100	2·950	2·790	3·100	3·000	2·820	2·640	3·000	2·900	2·670	2·510
175	3·940	3·780	3·610	3·450	3·760	3·630	3·450	3·280	3·630	3·510	3·300	3·100	3·510	3·380	3·120	2·950
200	4·500	4·340	4·140	3·960	4·320	4·170	3·960	3·760	4·170	4·010	3·780	3·530	4·010	3·860	3·580	3·380
225	5·050	4·880	4·650	4·470	4·850	4·670	4·470	4·220	4·670	4·520	4·240	3·990	4·520	4·340	4·040	3·780

Table 4.3.4. Ceiling joists

Dead load in kg/m^2 supported by the joists excluding the weight of the joists

Basic size of joist (mm)	Not more than 50				51 to 75 incl.				76 to 100 incl.				101 to 125 incl.			
	Spacing of joist in mm for centres up to															
	400	450	525	600	400	450	525	600	400	450	525	600	400	450	525	600
	Maximum span of joist in metres															
38 × 75	1·730	1·650	1·570	1·520	1·600	1·550	1·470	1·400	1·520	1·450	1·350	1·270	1·450	1·350	1·240	1·170
100	2·310	2·240	2·130	2·030	2·160	2·080	1·980	1·850	2·030	1·960	1·830	1·700	1·930	1·830	1·700	1·570
125	2·900	2·790	2·670	2·570	2·720	2·620	2·490	2·340	2·570	2·460	2·290	2·130	2·410	2·290	2·130	1·980
150	3·430	3·300	3·150	3·000	3·180	3·070	2·920	2·740	3·000	2·900	2·690	2·510	2·840	2·690	2·490	2·340
50 × 75	1·880	1·800	1·730	1·650	1·750	1·680	1·600	1·520	1·650	1·600	1·500	1·450	1·570	1·500	1·420	1·350
100	2·510	2·440	2·310	2·210	2·360	2·260	2·160	2·060	2·210	2·130	2·030	1·960	2·110	2·030	1·930	1·800
125	3·150	3·050	2·900	2·790	2·950	2·840	2·690	2·590	2·790	2·670	2·540	2·440	2·640	2·540	2·410	2·260
150	3·170	3·580	3·400	3·280	3·450	3·330	3·180	3·050	3·280	3·150	3·000	2·870	3·120	3·000	2·840	2·670
63 × 150	3·990	3·860	3·680	3·530	3·730	3·610	3·430	3·280	3·530	3·400	3·230	3·100	3·350	3·230	3·070	2·950
175	4·650	4·500	4·290	4·110	4·370	4·190	4·010	3·840	4·110	3·960	3·780	3·630	3·940	3·780	3·610	3·450
200	5·310	5·130	4·900	4·720	4·980	4·800	4·600	4·390	4·720	4·550	4·340	4·170	4·500	4·340	4·140	3·960
225	5·970	5·770	5·510	5·310	5·590	5·410	5·160	4·950	5·310	5·110	4·880	4·670	5·050	4·880	4·650	4·470

Table 4.3.5

Common or jack rafters for roof pitch over 35° but less than 40°

Dead load in kg/m² supported by rafters excluding the weight of the rafters	Not more than 50				51 to 75 incl.				76 to 100 incl.			
	Spacing of rafter in mm for centres up to											
Basic size of rafter (mm)	400	450	525	600	400	450	525	600	400	450	525	600
	Maximum span of rafter in metres											
38 × 75	2·160	2·060	1·930	1·800	1·900	1·830	1·750	1·630	1·730	1·680	1·600	1·500
100	2·900	2·770	2·590	2·410	2·570	2·460	2·340	2·180	2·340	2·240	2·130	2·010
125	3·630	3·480	3·230	3·050	3·200	3·070	2·920	2·740	2·920	2·820	2·690	2·540
50 × 75	2·360	2·260	2·160	2·060	2·080	2·010	1·900	1·830	1·880	1·830	1·730	1·650
100	3·150	3·020	2·900	2·740	2·770	2·678	2·540	2·440	2·540	2·440	2·340	2·240
125	3·910	3·780	3·610	3·450	3·480	3·350	3·200	3·070	3·180	3·070	2·920	2·790

Table 4.3.6 Purlins supporting rafters to which Table 4.3.5 relates for roofs having a pitch over 35° but less than 40°

Dead load in kg/m² supported by rafters as calculated in Table 3	Not more than 50					51 to 75 incl.					76 to 100 incl.				
	Spacing of purlins in metres														
	1·830	2·130	2·440	2·740	3·050	1·830	2·130	2·440	2·740	3·050	1·830	2·130	2·440	2·740	3·050
Basic size of purlin (mm)	Maximum span of purlin in metres														
50 × 100	1·400	1·300	1·220	1·140	1·090	1·270	1·190	1·120	1·040	0·990	1·170	1·090	1·020	0·940	0·860
125	1·780	1·630	1·520	1·450	1·370	1·570	1·500	1·400	1·320	1·240	1·450	1·370	1·300	1·190	1·070
150	2·080	1·930	1·800	1·700	1·630	1·880	1·750	1·650	1·550	1·470	1·700	1·630	1·520	1·400	1·270
175	2·440	2·260	2·110	2·010	1·900	2·180	2·060	1·930	1·830	1·730	2·010	1·900	1·780	1·650	1·470
63 × 150	2·260	2·160	2·030	1·900	1·830	2·010	1·930	1·830	1·750	1·650	1·850	1·750	1·680	1·600	1·520
175	2·67	2·540	2·390	2·240	2·130	2·360	2·260	2·160	2·030	1·930	2·160	2·060	1·980	1·880	1·800
200	3·050	2·900	2·720	2·570	2·440	2·720	2·590	2·460	2·340	2·240	2·490	2·360	2·260	2·160	2·060
75 × 175	2·280	2·690	2·590	2·460	2·340	2·510	2·390	2·290	2·210	2·130	2·310	2·210	2·110	2·030	1·960
200	3·230	3·070	2·950	2·820	2·690	2·900	2·740	2·640	2·540	2·440	2·640	2·510	2·410	2·310	2·240
225	3·630	3·480	3·330	3·200	3·020	3·250	3·100	2·970	2·840	2·770	3·000	2·840	2·720	2·620	2·540

Table 4.3.7 Purlins supporting sheeting or decking for roofs having a pitch of 22½° or more

Dead load in kg/m² supported by the purlin excluding the weight of the purlin

Basic size of purlin (mm)	Not more than 25					26 to 50 incl.					51 to 75 incl.				
	Spacing of purlins in metres														
	0·914	1·220	1·520	1·830	2·130	0·914	1·220	1·520	1·830	2·130	0·914	1·220	1·520	1·830	2·130
	Maximum span of purlin in metres														
50 × 75	1·350	1·170	1·040	0·940	0·890	1·190	1·040	0·910	0·610	0·700	1·090	0·940	0·840	0·760	0·710
100	1·700	1·470	1·320	1·190	1·120	1·500	1·320	1·170	1·170	0·990	1·370	1·190	1·070	0·960	0·890
125	2·010	1·750	1·570	1·420	1·320	1·800	1·550	1·400	1·270	1·170	1·630	1·420	1·270	1·170	1·070
150	2·260	1·980	1·750	1·630	1·500	2·030	1·750	1·570	1·450	1·320	1·830	1·600	1·420	1·300	1·220
175	2·540	2·210	1·980	1·800	1·680	2·260	1·960	1·750	1·600	1·500	2·060	1·780	1·600	1·470	1·330
200	2·770	2·410	2·160	1·980	1·830	2·490	2·160	1·930	1·780	1·650	2·260	1·960	1·750	1·600	1·500
63 × 150	2·690	2·360	2·110	1·930	1·780	2·410	2·110	1·880	1·730	1·600	2·210	1·900	1·700	1·570	1·450
175	3·020	2·640	2·360	2·160	2·010	2·720	2·360	2·110	1·930	1·800	2·460	2·160	1·930	1·750	1·630
200	3·330	2·920	2·620	2·350	2·210	3·000	2·570	2·340	2·130	1·980	2·720	2·368	2·130	1·900	1·800
75 × 175	3·480	3·050	2·740	2·510	2·310	3·120	2·720	2·440	2·240	2·080	2·840	2·430	2·240	2·030	1·880
200	3·840	3·320	2·020	2·770	2·570	3·450	3·000	2·690	2·460	2·290	2·030	3·150	2·740	2·460	2·080
225	4·090	3·660	3·330	3·020	2·790	3·760	3·280	2·950	2·690	2·490	3·430	3·000	2·690	2·460	2·290

Section 5
Loadbearing Brickwork and Blockwork

5. Loadbearing Brickwork and Blockwork

Relevant code CP 111, 'Structural Recommendations for Loadbearing Walls.'

The notes in this section apply to the following materials solidly put together with mortar: Bricks and suitable blocks of clay, concrete, natural or cast stone.

5.1 SAFE LOADS

The safe load for a wall, pier, or column, of bricks or blocks depends on many factors such as:

1. The strength of the individual units.
2. The strength of the mortar.
3. The workmanship.
4. The slenderness ratio which involves the thickness and height of a member and the conditions of support to the top and bottom of the wall, pier or column.
5. The disposition of the load, e.g. whether axial or eccentric.

Definitions

$$\text{Slenderness ratio} = \frac{\text{Effective height}}{\text{Effective thickness}}$$

Effective Height of Walls

When the floor or roof spans at right angles to the wall with sufficient bearing and anchorages,

Effective height = $^3/_4$ of actual height between centres of supports

For concrete floors having a bearing on walls, irrespective of the direction of span,

Effective height = $\frac{3}{4}$ of actual height

For floors or roof spanning parallel with the wall without bearing,

Effective height = actual height

(Note that metal ties anchored as for example into concrete blocks built into the wall should be provided. See CP 111).

For walls with no lateral support at the top,

Effective height = $1\frac{1}{2}$ times actual height

5.2 EFFECTIVE THICKNESS

Solid Walls

Figure 5.2.1

Effective thickness = t_W = actual thickness of wall (A)

Cavity Walls (with adequate ties)

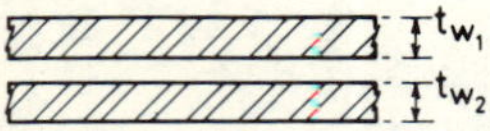

Figure 5.2.2

Effective thickness = $\frac{2}{3}(t_{W_1} + t_{W_2})$(B)

Thus, e.g., if t_{W_1} = 112·5 mm and t_{W_2} = 112·5 mm:

Effective thickness = $\frac{2}{3}$ (112·5 + 112·5) = 150 mm

This applies when the load is carried by both leaves. Where the load is carried by one leaf only the permissible stress for that leaf may be based on the slenderness ratio calculated from an effective thickness of $\frac{2}{3}(t_{W_1} + t_{W_2})$. This allows for the other leaf.

Alternatively the effective thickness may be taken as the actual thickness of the loadbearing leaf. In either alternative, only the actual thickness of the loadbearing leaf should be used in calculating the permissible load.

Walls Stiffened by Intersecting Walls or Piers

A pier is a member which forms an integral part of a wall.

The effective height of the stiffened wall is as given for un-stiffened walls.

Effective Thickness

The effective thickness of the stiffened wall as given in (A) above for solid walls may be multiplied by the appropriate stiffening coefficient given in Table 5.2.1.

Table 5.2.1 Stiffening Coefficient (Solid Walls)

$\frac{\text{Pier spacing (centre to centre)}}{\text{Pier width}}$ i.e. $\frac{s}{b}$	$\frac{\text{Pier thickness}}{\text{Effective wall thickness}}$ 1	2	3
6	1·0	1·4	2·0
10	1·0	1·2	1·4
20	1·0	1·0	1·0

Intermediate values can be obtained by linear interpolation.

Example 1. Solid Wall

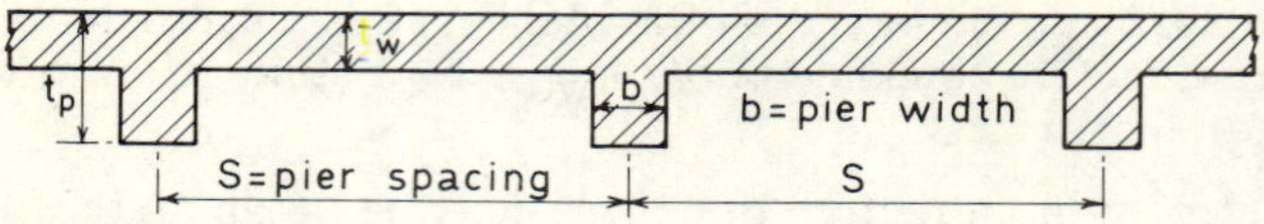

Figure 5.2.2

If $t_w = 225$ mm, $t_p = 450$ mm, $b = 450$ mm, $s = 4{\cdot}5$ m,

$$\frac{s}{b} = \frac{4{\cdot}500}{0{\cdot}450} = 10; \ \frac{t_p}{t_w} = \frac{450}{225} = 2$$

From Table 5.2.1 coefficient = 1·2.

Effective thickness of wall for calculation of slenderness ratio $= 1{\cdot}2 \times t_w = 1{\cdot}2 \times 225 = 270$ mm

Example 2. Cavity Wall

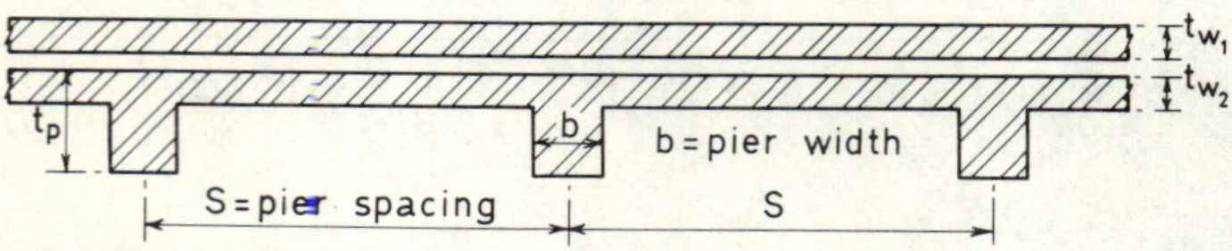

Figure 5.2.3

Calculate t_p/t_{w_2} and s/b; obtain the stiffening coefficient from Table 5.2.1 and multiply t_{w_2} by the coefficient to give an effective thickness t_e.

Then, effective thickness of the stiffened wall $= \frac{2}{3}(t_e + t_w)$.

Thus, e.g., if $t_{w_1} = 112{\cdot}5$ mm, $t_{w_2} = 112{\cdot}5$ mm, $t_p = 225$ mm, $b = 450$ mm, $s \doteq 4{\cdot}5$ m,

$$\frac{t_p}{t_{w_2}} = \frac{225}{112{\cdot}5} = 2 \quad \frac{s}{b} = \frac{4{\cdot}500}{0{\cdot}450} = 10$$

From Table 5.2.1 coefficient = 1·2.

$$t_e = 1{\cdot}2 \times 112{\cdot}5 = 135 \text{ mm}$$

Effective thickness of wall for calculation of slenderness ratio

$$\frac{2}{3}(135 + 112{\cdot}5) = 165 \text{ mm}$$

Intersecting Walls

When a wall is stiffened by intersecting walls instead of piers, assume the intersecting wall to be a pier where b = the thickness of the intersecting wall and t_p = 3 times the thickness of the stiffened wall. The stiffening coefficients may then be obtained as described above.

Columns

A column is an isolated loadbearing member whose width is not more than four times its thickness (Figure 5.2.4).

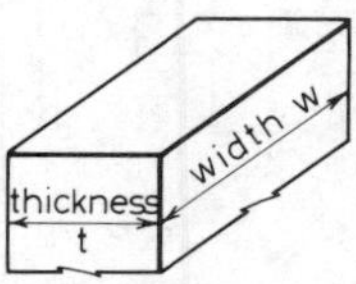

Figure 5.2.4

Effective Height of Columns

CP 111 states: Where a column is provided with lateral support parallel to the line of one of the horizontal surface dimensions at the top, its effective height relative to the direction of the top support should be taken as the height between supports; its effective height at right angles to this should be taken as twice its height above the lower support. In the absence of top support, its effective height relative to both directions should be taken as twice its height above the lower support.

The G.L.C. By-laws state: The effective height of a storey height of a pier (the word column is not mentioned in the section of the By-laws dealing with "Walls and Piers").

(a) with lateral support at the top of that storey height, shall be the storey height;

(b) without lateral support at the top of that storey height, shall be twice the storey height.

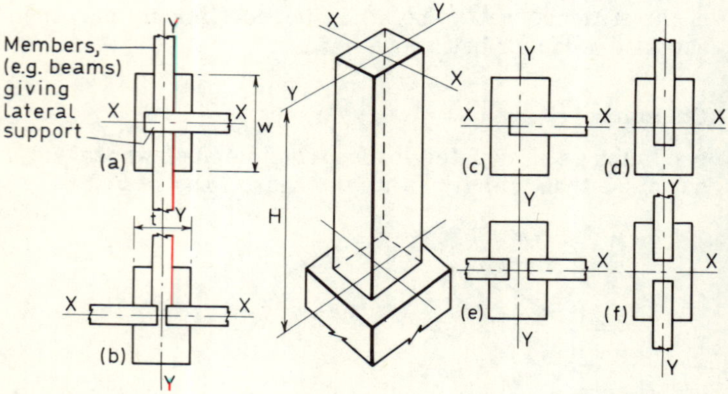

Figure 5.2.5

The G.L.C. By-laws further state that the slenderness ratio of any storey height of a pier shall be the ratio of the effective height to the horizontal dimension lying in the direction of the lateral support determining such storey height.

The author's interpretation of these clauses is given below with reference to Figure 5.2.5 (Table 5.2.2).

Table 5.2.2

H = actual or storey height of the column; l = effective height; t = thickness of column, i.e. smaller dimension of cross-section; w = width of column, i.e. larger dimension of cross-section.

	Effective height l		
Fig. 5.2.5	axis XX	axis YY	Slenderness ratio
a	H	H	H/t
b	H	H	H/t
c	2H	H	the greater of 2H/w and H/t
d	H	2H	2H/t
e	2H	H	the greater of 2H/w and H/t
f	H	2H	2H/t

Permissible Stresses for Brickwork and Blockwork

Table 5.2.3 gives the basic permissible stresses (for a slenderness ratio not exceeding 6) depending on the crushing strength of the structural unit (brick or block) and the grade of mortar.

Table 5.2.3 Basic compressive stresses for brickwork or blockwork (at or after stated times)

Description of mortar	Mix (parts by volume)			Hardening time after completion of work†	Basic stresses (N/mm²) corresponding to units whose crushing strength (N/mm²) is:								
	Cement	Lime	Sand		2·8	7	10·5	21	28	35	52·5	70	98 or more
				days									
Cement	1	0–¼*	3	7	0·28	0·7	1·05	1·68	2·1	2·52	3·57	4·62	5·95
	1	½	4½	14	0·28	0·7	0·98	1·47	1·75	2·10	2·87	3·64	4·55
Cement–lime	1	1	6										
Cement with plasticiser	1	—	6	14	0·28	0·7	0·98	1·33	1·61	1·89	2·52	3·15	3·85
Masonry cement	—	—	—										
Cement–lime	1	2	9										
Cement with plasticiser	1	—	8	14	0·28	0·56	0·84	1·19	1·47	1·68	2·10	2·52	3·15
Masonry cement	—	—	—										
Cement–lime	1	3	12	14	0·21	0·49	0·70	0·98	1·19	1·40	1·75	2·10	2·45
Hydraulic lime	—	1	2	14	0·21	0·49	0·70	0·98	1·19	1·40	1·75	2·10	2·45
Non-hydraulic	—	1	3	28	0·21	0·42	0·56	0·70	0·77	0·84	1·05	1·19	1·40

* The inclusion of lime in cement mortars is optional.

† These periods should be increased by the full amount of any time during which the air temperature remains below 4·4 °C and half the time during which the temperature is between 4·4 °C and 10 °C.

Where the cross-sectional plan of a wall or column does not exceed 0·324 m^2 the basic stress should be multiplied by a reduction factor equal to $0{\cdot}75 + A/1{\cdot}296$ where A is the area (m^2) of the horizontal cross-section of the wall or column.

Table 5.2.4 is given merely as an approximate guide to the crushing strengths of a few types of bricks, etc. Manufacturers should be consulted for more accurate values.

Table 5.2.4 Guide to Crushing Strengths

Type	Crushing strength (N/mm^2)
Engineering bricks Grade A	70
Grade B	49
Sand lime bricks	7-21 depending on type
Concrete bricks	7-17·5 depending on type
Staffordshire Blue	112
Accrington Red	70
Southwater (Sussex)	35
Leicester Red wirecuts	28
London Stocks	3·5-10·5
Flettons	17·5-31·5
Limestone	14-56
Sandstone	about 70
Granite	70-140

Axially Loaded Members

Permissible stress = basic stress from Table 5.2.3 multiplied by the reduction factor for slenderness given in Table 5.2.5

Table 5.2.5 Reduction Factors for Slenderness

	Reduction factor			
		Eccentricity of vertical loading as a proportion of the thickness of the member		
Slenderness ratio	Axially loaded	1/6	1/3	1/2
6	1·00	1·00	0·98	0·97
8	0·92	0·92	0·87	0·85
10	0·84	0·82	0·76	0·73
12	0·76	0·72	0·65	0·60
14	0·67	0·62	0·54	0·48
16	0·58	0·52	0·42	0·35
18	0·50	0·42	0·30	0·23
21	0·47	0·38	0·25	0·17
24	0·44	0·34	0·19	0·11

Linear interpolation between values for the reduction factor is permissible.

Walls Subjected to Eccentric Loading and/or Lateral Forces

The maximum calculated stress (i.e. axial or direct stress

plus bending stress) may exceed the values of the stress arrived at by the use of Table 5. 2. 5 by not more than 25% provided that such excess is due solely to eccentricity and/or lateral forces.

Maximum Slenderness Ratios of Walls

The slenderness ratio for unreinforced walls of brickwork or blockwork should not exceed:

With lime mortar, in buldings of more than 2 storeys: 12
With lime mortar, in buildings of not more than 2 storeys: 18
With mortar other than lime mortar, more than 2 storeys: 18
With mortar other than lime mortar, in dwellings of not more than 2 storeys: 24

Walls Subjected to Concentrated Loads

For example, at girder bearings, column bases, lintels

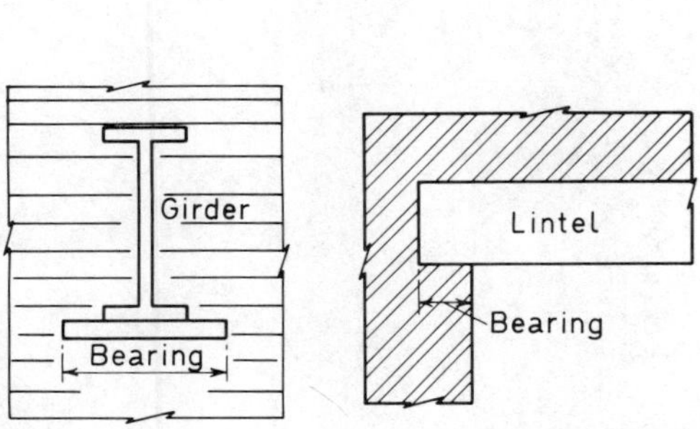

Figure 5. 2. 6

Referring to Figure 5. 2. 6, the additional stress on the area immediately under the girder or lintel bearing is to be calculated. Let this stress be f_1. If the stress due to other causes, e.g. weight of brickwork above etc. = f_2, then the maximum stress = $f_1 + f_2$. This stress can exceed the permissible axial stress given in Table 71 by not more than 50%.

Reduction of Stress due to Slenderness of Member

The reduction factor for slenderness ratios need not be used for sections within one-eighth of the height of the member above or below the positions of the lateral supports.

5.3 DISPERSION OF LOAD

The angle of dispersion of loading of walls should be taken as not more than 45° from the direction of such loading (Figure 5.3.1).

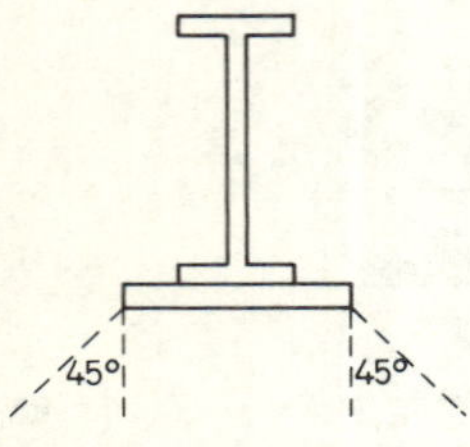

Figure 5.3.1

Permissible Shear Stress in Brickwork or Blockwork

In the case of brick walls resisting horizontal forces in the plane of the wall, the permissible shear stress calculated on the area of the horizontal mortar bed joint should be taken as

0·1 N/mm^2 for walls built with a 1 : 1 : 6 mortar mix (or its equivalent) and 0·14 N/mm^2 when a 1 : ¼ : 3 for stronger mortar is used; or one third of the compressive stress produced by dead load at the level under consideration. The permissible stress should be the greater of the two values specified for each mortar but should not exceed 0·21 N/mm^2.

Section 6
Retaining Walls

Vertical Wall and Horizontal Ground Design
Reinforced Concrete Cantilever Walls

6. Retaining Walls

Relevant code: Civil Engineering Code of Practice No. 2 'Earth Retaining Structures.'

Pressures on Retaining Walls

Cohesionless soils, e.g. sands and gravels.

The pressures given in this book apply to cohesionless soils only. Cohesive soils, e.g. clays and silts, are too complicated to be dealt with in a work of this description and reference should be made to the Code of Practice.

For cohesionless soils, the active pressure behind a wall depends on: the density w of the retained material, the angle of internal friction of the soil ϕ and the angle of friction δ of the earth on the wall back.

6.1 VERTICAL WALL AND HORIZONTAL GROUND

The pressure p in kN/m^2 at any depth h below the ground surface is $p = Kwh \sec \delta$ and the total active pressure on one metre run of wall (P) is $K\dfrac{9{\cdot}81wH^2}{2} \sec \delta kN$

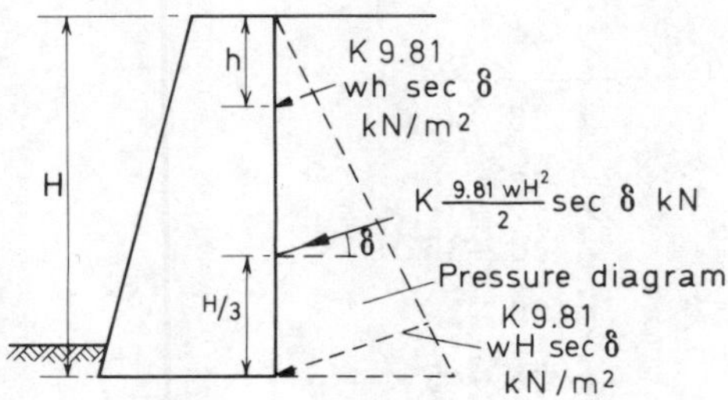

Figure 6.1.1

Where no test data are available δ may be taken as 20° for walls of concrete or brick.

K is obtained from Table 6.1.1.

Table 6.1.1 Values of K for Cohesionless Materials, Vertical Walls with Horizontal Ground

Values of δ	Values of ϕ 25°	30°	35°	40°	45°
	Values of K				
0	0·41	0·33	0·27	0·22	0·17
10	0·37	0·31	0·25	0·20	0·16
20	0·34	0·28	0·23	0·19	0·15
30	—	0·26	0·21	0·17	0·14

Typical values of ϕ (angle of internal friction) and w (weight per cubic foot of moist soils) are given in Tables 6.1.2 and 6.1.3.

Table 6.1.2 Typical Values of ϕ for Cohesionless Materials

Materials	ϕ
Sandy gravel	35°–45°
Compact sand	35°–40°
Loose sand	30°–35°
Shale filling	30°–35°
Rock filling	35°–45°
Ashes or broken brick	35°–45°

Table 6.1.3 Densities of cohesionless materials (kg/m^3)

Material		Density when drained above ground water level (w)	Density when submerged below ground water level (w_s)
Gravel		1600–2000	960–1280
Coarse and medium sands		1680–2080	960–1280
Fine and silty sands		1760–2160	960–1280
Rocks	granites and shales	1600–2080	960–1280
Rocks	basalts and dolerites	1760–2240	1120–1600
Rocks	limestones and sandstones	1280–1920	640–1280
Rocks	chalk	960–1280	320–640
Broken brick		1120–1760	640–960
Ashes		640–960	320–480

Rankine's Formula for Horizontal Ground

This assumes that the angle of friction (δ) of the earth on the wall back is zero so that the total pressure on the back of a vertical wall is horizontal.

In Figure 6.1.2 it may be noted that $(1 - \sin \phi)/(1 + \sin \phi)$ will equal the values of K in Table 6.1.1 when δ is taken as zero.

p = pressure in kN/m^2 at a depth H metres.

P = total pressure in kN per metre run of wall.

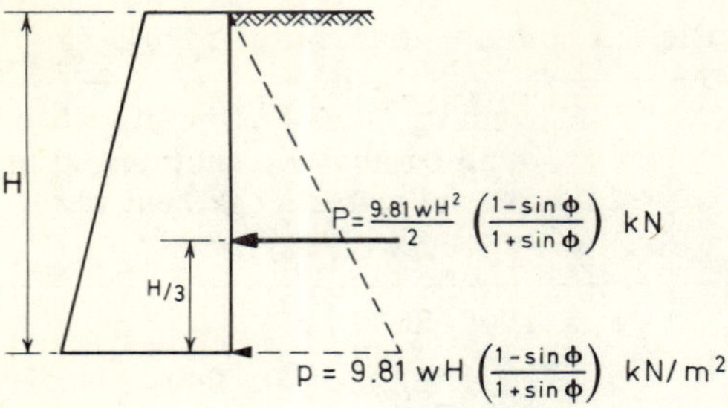

Figure 6.1.2

Rankine's formula for ground sloping at an angle equal to the angle of repose

It is assumed that the angle of repose is equal to the internal angle of friction (ϕ).

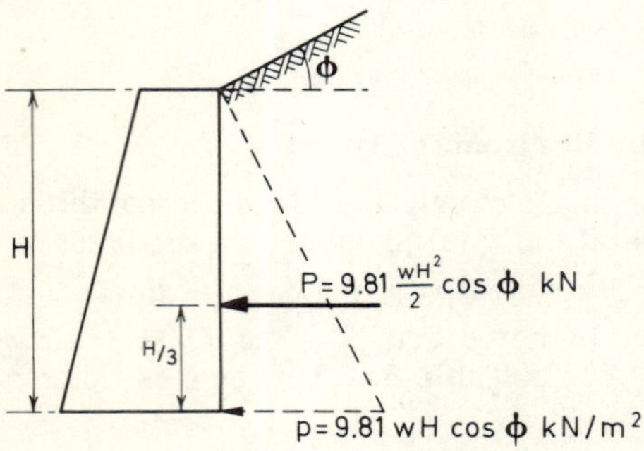

Figure 6.1.3

Rankine's formula for horizontal ground and superimposed load of W kN/m² due to traffic or buildings, etc.

Replace the load of W kN/m² by equivalent height (h m) of earth where $h = W/9{\cdot}81w$ (w = density of soil in kg/m³) (Figure 6.1.4).

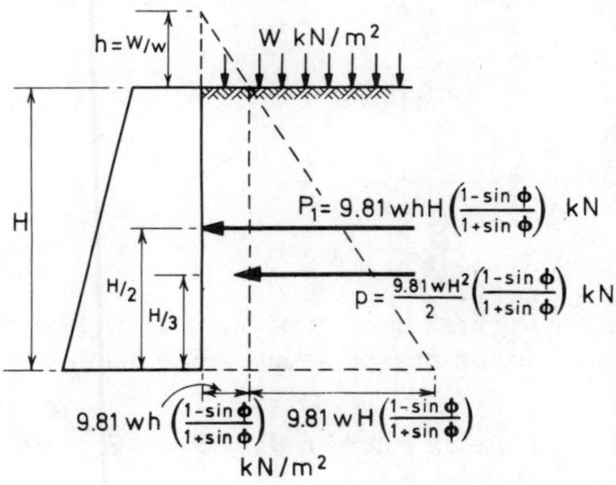

Figure 6.1.4

Total pressure per metre run of wall = $P + P_1$

Pressures Due to Saturated Soils

Assume $\delta = 0$.

Referring to Figure 6.1.5 the pressure in kN/m² at any depth equals the pressure due to the normal density of the soil (say the moist density) above the ground water level, plus the

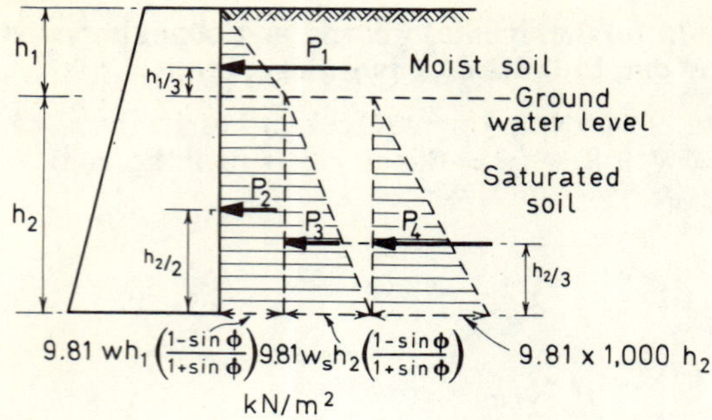

Figure 6.1.5

pressure due to the saturated density of the soil (w_s) below the ground water level, plus the pressure due to the density of the water (1000 kg/m³).

The total pressure per metre run of wall = $P_1 + P_2 + P_3 + P_4$ where

$$P_1 = \frac{9{\cdot}81 w h_1^2}{2}\left(\frac{1 - \sin\phi}{1 + \sin\phi}\right) \text{kN}$$

$$P_2 = 9{\cdot}81 w h_1 h_2 \left(\frac{1 - \sin\phi}{1 + \sin\phi}\right) \text{kN}$$

$$P_3 = \frac{9{\cdot}81 w_s h_2^2}{2}\left(\frac{1 - \sin\phi}{1 + \sin\phi}\right) \text{kN}$$

$$P_4 = \frac{9{\cdot}81 \times 1000 h_2^2}{2} \text{kN}$$

Lateral Pressure on Reinforced Concrete Walls

For the design of the stem of the wall and for the calculation of the forces on the base, the lateral pressure should be taken as acting horizontally. (i.e. $\delta = 0$).

See Figure 6.1.6.

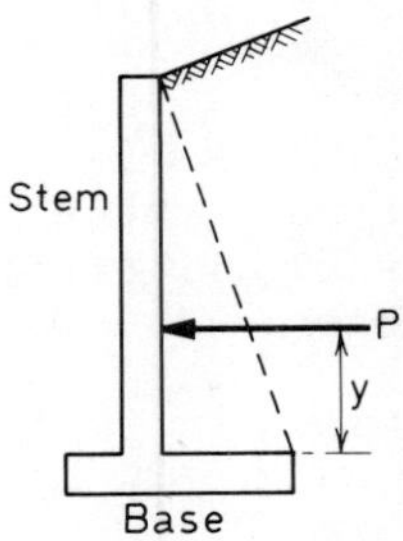

Figure 6.1.6

The stem can be designed as a slab for a moment of Py.

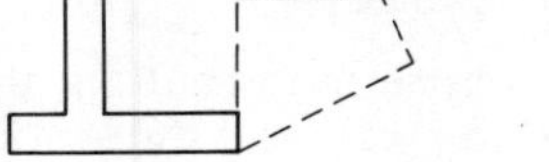

Figure 6.1.7

When the wall is on the point of failure by sliding or tilting forward P can be taken as acting as in Figure 6.1.7 where δ (the angle of friction on the back of the wall) is taken as ϕ (the angle of internal friction of the backing).

6.2 DESIGN

Gravity Walls

1. Assume a cross-section for the wall (width of base somewhere near half the height of the wall) and calculate the vertical line in which the centroid lies (by taking moments of areas about the back of the wall) (Figure 6.2.1 (a) and (b)).

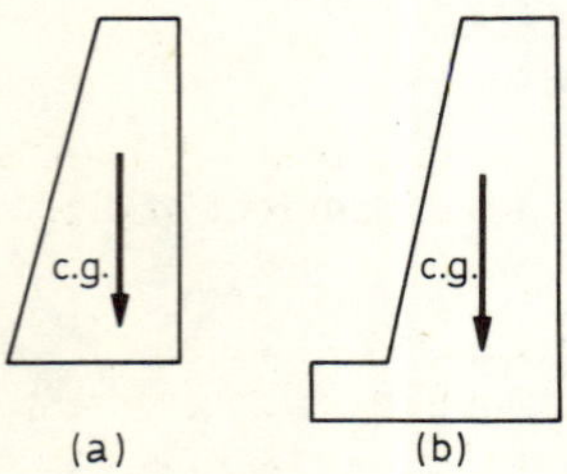

Figure 6.2.1

2. Calculate the weight of one metre of wall (W) and the total pressure (P) due to the earth backing.
3. Find, graphically or otherwise where the resultant thrust (R) of P and W cuts the base of the wall and determine the eccentricity e (Figure 6.2.2)

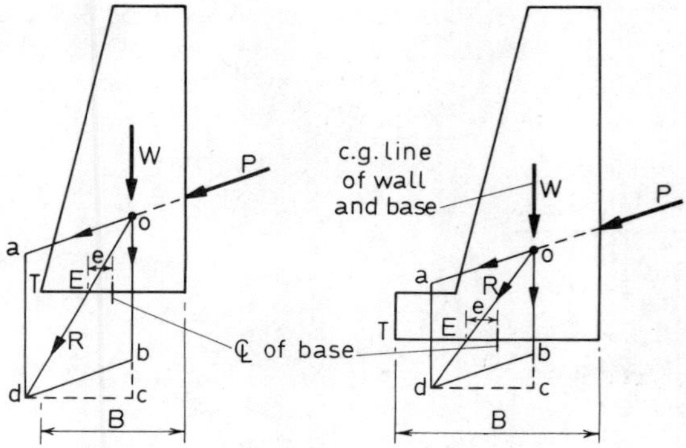

Figure 6.2.2

In gravity walls, the resultant thrust R should not fall outside the middle third of the base, i.e. TE should be at least equal to B/3.

In Figure 6.2.2 to scale, 0a represents P, 0b represents W, 0c is the vertical component R_V of the resultant R, and cd is the horizontal component R_H of R. When P is horizontal, the value of the eccentricity e can be found simply by calculation as shown in Figure 6.2.3.

$$x = \frac{Py}{W}$$

$$e = x - u$$

where u is the distance from the centre of gravity of the wall to the mid point of the base.

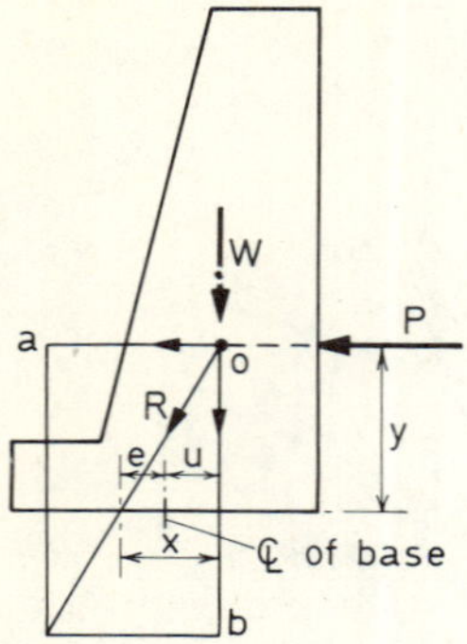

Figure 6.2.3

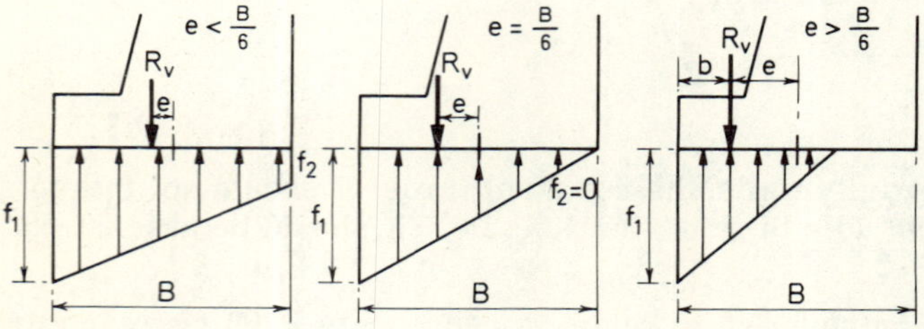

Figure 6.2.4

4. Calculate the bearing pressure on the foundation soil (Figure 6.2.4).

(a) When R falls inside the middle third (e is less than B/6),

$$\text{Maximum pressure } f_1 = \frac{R_V}{B}\left(1 + \frac{6e}{B}\right) \text{ kN/m}^2$$

$$\text{Minimum pressure } f_2 = \frac{R_V}{B}\left(1 - \frac{6e}{B}\right) \text{ kN/m}^2$$

(b) When R falls at the middle third ($e = B/6$)

$$f_1 = \frac{2R_V}{B}. \quad f_2 = 0$$

(c) When R falls outside the middle third (not recommended) (e is greater than B/6), since no tensile stresses are possible,

$$f_1 = \frac{2}{3}\frac{R_V}{b}$$

The maximum pressure f_1 in all cases should not exceed the maximum safe bearing capacity of the soil. A guide is given in Table 6.2.1.

Table 6.2.1 Maximum Safe Bearing Capacities for Foundations of Width Greater than 1 m and depth not less than 600 mm Cohesionless Soils

Type of soil	Typical values of ϕ (degrees)	Max. safe bearing capacity (kN/m^2) Dry*	Submerged
Compact well-graded sands and gravel–sand mixtures	40–45	400–600	200–300
Loose well-graded sands and gravel–sand mixtures	35–40	200–400	100–200
Compact uniform sands	35–40	200–400	100–200
Loose uniform sands	30–35	100–200	50–100

* Dry means that the ground water level is at a depth not less than the width of the foundation B below the bottom of the foundation.

5. Check for resistance to Sliding

Resistance to sliding is given by the frictional resistance of the soil under the base plus the passive resistance (if any) of the soil in front of the wall.

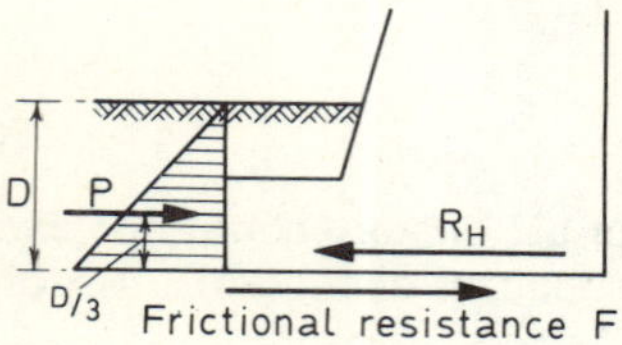

Figure 6.2.5

In Figure 6.2.5 R_H is the force causing sliding = horizontal component of resultant R

F is the frictional resistance of the soil. P is the passive resistance of the soil. R_V is the vertical component of the resultant R. There should be a factor of safety of about 2 against sliding, therefore

$$\frac{F + P}{R_H} \text{ should} = 2$$

F may be taken as $R_V \tan \phi$ for a concrete foundation cast in situ.

ϕ = angle of internal friction of the soil

δ = angle of wall friction

P may be taken as

$$\frac{9{\cdot}81wD^2}{2}\left(\frac{1 + \sin \phi}{1 - \sin \phi}\right)$$

where w should preferably be taken as the dry density of the soil. (Note the plus sign in the numerator when dealing with passive pressure.)

6.3 REINFORCED CONCRETE CANTILEVER WALLS

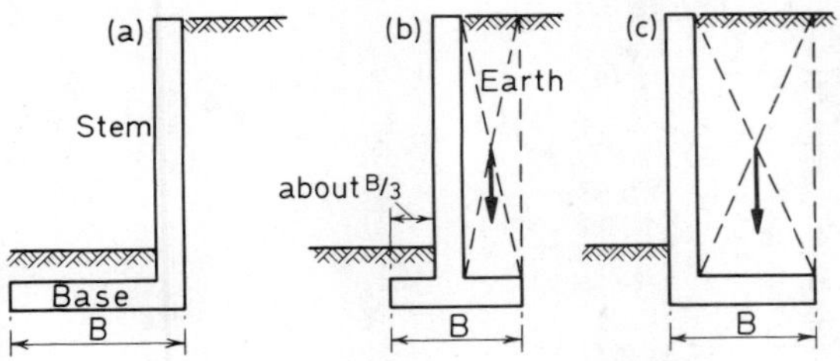

Figure 6.3.1

Design

1. Calculate the lateral pressure (P) per metre run of wall, calculate the maximum bending moment and obtain the thickness of the stem and the amount of steel. As a rough guide the thickness of the stem where it joins the base is one tenth of the height of the stem.

2. Assume dimensions for the base. As a rough guide the thickness of the base should be at least equal to the thickness of the stem; the width B of the base is about $2/3$ the height of the wall for wall (a) Figure 6.3.1 and about $1/2$ the height of the wall for walls (b) and (c).

3. Calculate the weight per metre run of the wall and the weight per metre run of the earth over the base (in walls (b) and (c)). Calculate the position of the vertical line in

which the centre of gravity of the wall and earth lies. An example is given in Figure 6.3.2.

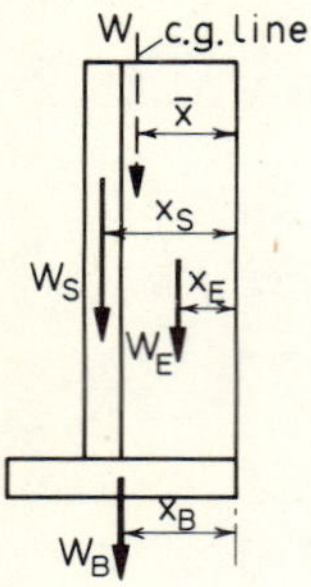

Figure 6.3.2

W_S = weight of stem per metre run (kN).

W_B = weight of base per metre run (kN).

W_E = weight of earth above base (kN).

Taking moments about the heel

$$(W_S + W_B + W_E)\,\bar{x} = W_S x_S + W_B x_B + W_E x_E$$

Total vertical weight $= W = W_S + W_B + W_E$

4. Find where the resultant of W and P cuts the base (as for gravity walls).

5. Calculate the bearing pressures on the soil (as for gravity walls).

6. Check for resistance to sliding (as for gravity walls).

7. Calculate bending moments in base and design the steel, e.g. bending moment at XX Figure 6.3.3 is due to downward pressure of earth, downward pressure due to weight of base, and upward pressure due to reaction from soil.

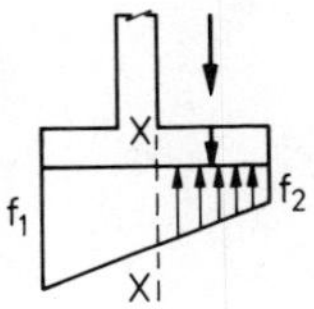

Figure 6.3.3

Section 7
Prestressed Concrete

7. Prestressed Concrete

7.1 INTRODUCTION

The syllabuses for many architectural and building courses require a mainly descriptive treatment of prestressed concrete. The basic principles of prestressing are very simple, though practical design examples are generally too complicated for abbreviated treatment and only the simplest case of beam design is considered.

Basic principles

Prestressing makes use of the high compressive strength of concrete by using high-tensile steel to introduce a compressive stress into the concrete before a load is applied, rather than, as in reinforced concrete, using steel to take the tensile stresses induced under load.

Prestressed members may be pre-tensioned, in which case the wires or cables are first stressed and the concrete cast around them; or post-tensioned in which case members are first cast and then the wires are passed through preformed ducts, stressed, and anchored at each end of the unit. Pre-tensioning is suited to factory production of small uniform units such as floor beams. Post-tensioning is used for larger structures such as bridges.

7.2 REQUIREMENTS OF CP115, 'THE STRUCTURAL USE OF PRESTRESSED CONCRETE IN BUILDING'.

Concrete

In order to take maximum advantage of prestressing, high-strength designed mixes are invariably specified. The required minimum works cube strength is 40 N/mm^2 for pretensioned work and 30 N/mm^2 for post-tensioning.

Table 7. 2. 1 Permissible compressive stresses in concrete. u_w = specified works cube strength at 28 days, u_t = specified minimum cube strength at transfer of load from steel to concrete.

Nature of loading	Permissible concrete stress (N/mm^2)
Maximum working load in bending	$0{\cdot}33u_w$ (for simply supported beams)
Transfer	$0{\cdot}50u_t$ (for triangular prestress distribution) $0{\cdot}40u_t$ (for uniform prestress distribution) or 20 N/mm^2 whichever is lower

Small tensile stresses, up to about 2 N/mm^2 may be allowed in concrete in buildings, though not in highway bridges.

Steel

(a) The initial stress in hard drawn steel wire to B.S. 2691 should not exceed 70 per cent of its tensile strength.

(b) Alloy steel bars should not carry an initial tensile stress exceeding 70 per cent of the ultimate strength of 85 per cent of the 0·2 per cent proof stress whichever is less. With pre-tensioned bars, initial tensile stress should not exceed 75 per cent of the ultimate strength or 90 per cent of the 0·2 per cent proof stress whichever is less.

Loss of prestress

There are a number of ways in which the prestressing force exerted by the jack is not transmitted effectively to the concrete, and the Code lists the following, with methods of calculating the loss of prestress:

1. Creep of steel and of concrete.
2. Elastic deformation of concrete.
3. Shrinkage of concrete.
4. At anchorage.
5. Steam curing.
6. Friction.

Cover

Cover required is 25 mm for inside work and 40 mm for external work. For factory made, pre-tensioned units, these values may be reduced to 12 mm and 25 mm.

Effective span

The effective span is the lesser of (a) the distance between centres of supports and (b) clear distance between supports plus the depth of the beam.

7.3 DESIGN OF SYMMETRICAL SECTION FOR SIMPLY SUPPORTED BEAM

Stresses in a prestressed beam at various stages of loading are shown in Figures 7.3.1 to 7.3.3.

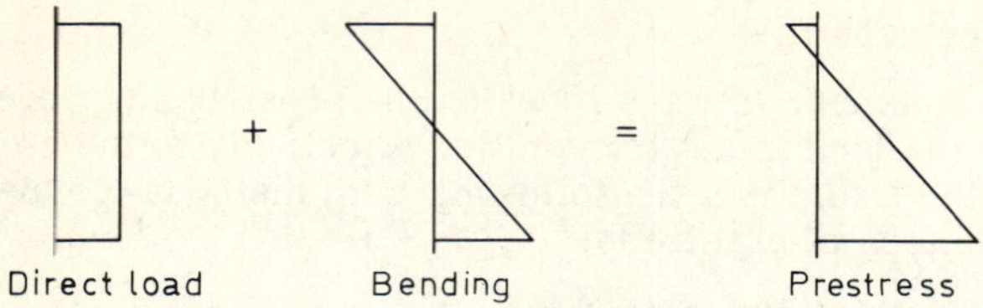

Figure 7.3.1

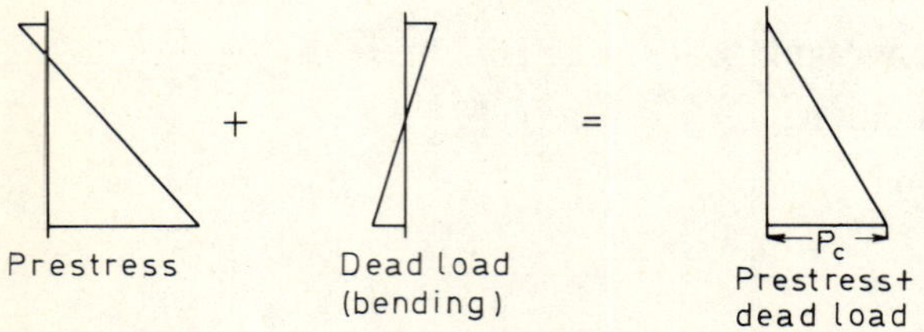

Figure 7.3.2

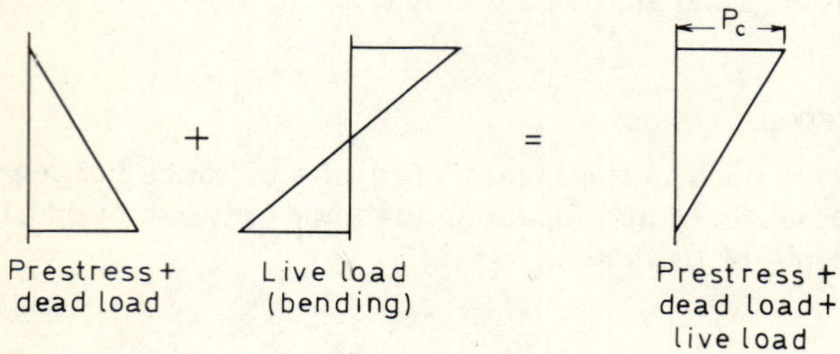

Figure 7.3.3

Symbols

P = prestressing force (N)
A = area of section (mm^2)

p_c = permissible maximum working compressive stress in concrete (N/mm^2)
M_D = Maximum bending moment due to weight of beam (N mm)
M_L = Maximum bending moment due to additional loads (N mm)
e = eccentricity of prestressing force (mm)
Z = section modulus (mm^3)

Design

Then an analysis of the conditions shown in Figures 7.3.1, 7.3.2, and 7.3.3. gives the following formulae:

$$P = \frac{p_c A}{2} \qquad \text{——— ①}$$

$$Z = \frac{M_L}{p_c} \qquad \text{——— ②}$$

$$e = \frac{M_D}{P} + \frac{M_L}{2P} \qquad \text{——— ③}$$

These three formulae enable a beam section to be designed with the appropriate prestressing force and eccentricity.

Section 8
Guide to Sizes

Roof Beams, Floor Beams, Columns, Bases
General Guide

8. Guide to Sizes

8.1 ROOF BEAMS, FLOOR BEAMS, COLUMNS, BASES

Architectural students frequently have to show on their drawings reasonable dimensions of structural members without having made calculations for them. It is not possible to give 'rules of thumb' for all types of structure. Certain information has already been given in other sections of the book. For example, Table 3.1.6 gives a useful guide to the thickness of slabs and beams. Table 2.1.2 can be used to give some idea of the minimum depth of steel beams.

Columns in multi-storey buildings are more difficult. The size of a steel column depends on its effective height, the area of each floor it supports, the number of storeys, the disposition of the load whether axial or eccentric, and so on.

The size of a reinforced concrete column can vary within fairly wide limits according to the proportions of the mix, the quality of the concrete, the percentage of steel, etc.

An attempt is made in three examples which follow to give beam, column, and foundation sizes for different types and dimensions of floor plans. In all examples, the building is of 12 storeys, i.e. ground floor to 11th floor plus a flat roof. Office loading including allowance for light partitions (2·5 kN/m^2) is assumed so that the sizes given will also be adequate for domestic flats. The roof and floors are assumed to be solid reinforced concrete, the columns and reinforcement are of mild steel and 200 kN/m^2 has been taken as the permissible bearing capacity of the soil. The columns given in the examples are properly cased in concrete with 40 mm cover and should be suitable for effective heights up to 4 metres.

It is strongly emphasized that all dimensions are approximate, and for any building which is to be erected, proper and detailed calculations should be made. Individual foundations for columns in a multi-storey building will be large and in an actual project it is possible that piles would be used. The sizes given for columns assume that the loads are axial. 'Wall' and corner columns will be subjected to eccentric loading but as a rule these columns are less heavily loaded than internal columns. Further, the bottom lengths of columns in a multi-storey building approximate to the axially loaded condition so the sizes given can be taken as approximately suitable for 'wall' and corner columns.

Mix proportions and steel percentages have been assumed in determining the approximate sizes of the reinforced concrete columns. Different proportions and percentages can be used and these will give, of course, different sizes.

Bearing in mind all the restrictions, the three examples give reasonable sizes of columns and foundations for ANY number of storeys up to 12. For different floor plans it may be useful to note that in Example 1 the column supports at each floor an area of 53·5 m^2, in Example 2, the area is 30 m^2 and the area of Example 3 is 26 m^2.

(In the case of steel sections, the existing size is given, with the dimensions in millimetres and kilogrammes per metre in parenthesis).

Example 1

In Fig. 8.1.1, S = secondary beam, M = main beam, C = axially loaded column, U.B. = universal beam, and U.C. = universal column.

Slab Thickness = 115 mm for roof and all floors.

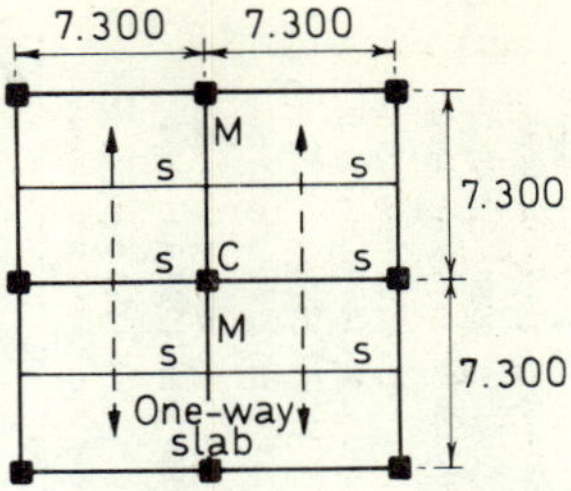

Figure 8.1.1

Roof beam S:
Steel. 15 in × 6 in × 45 lb U.B. (389 × 154 mm × 67 kg).
Reinforced concrete. Rib = 300 × 150 mm or 300 × 200 mm, i.e. 420 × 150 mm or 420 × 200 mm overall.

Roof beam M:
Steel. 21 in × 8¼ in × 68 lb U.B. (537 × 210 mm × 101 kg).
Reinforced concrete. Rib = 380 × 200 mm, i.e. 500 × 200 mm overall.

Floor beam S:
Steel. 16 in × 6 in × 50 lb U.B. (416 × 154 mm × 74 kg).
Reinforced concrete. Rib = 380 × 200 mm, i.e. 500 × 200 mm overall.

Floor beam M:
Steel 21 in × 8¼ in × 82 lb U.B. (545 × 212 mm × 122 kg).
Reinforced concrete. Rib = 460 × 230 mm i.e. 570 × 230 mm overall.

Example 1 Figure 8.1.1	Cased steel column (C) below	R.C. column (C) below	R.C. Foundation
Roof	6 in × 6 in × 15·7 lb U.C. (152 × 152 mm × 23 kg) 250 × 250 mm overall	1·2·4 mix. 0·8% steel 250 × 250 mm	1370 mm square 300 mm deep
11th Floor	8 × 8 in × 31 lb U.C. (203 × 203 mm × 46 kg) 300 × 300 mm overall	1·2·4 mix. 1% steel 370 mm square or 300 × 450 mm	2060 mm square 510 mm deep
10	8 × 8 in × 40 lb U.C. (210 × 205 mm × 59 kg) 310 × 310 mm overall	1·2·4 mix. 2% steel 410 mm square or 300 × 560 mm	2520 mm square 610 mm deep
9	8 × 8 in × 58 lb U.C. (222 × 209 mm × 86 kg) 330 × 310 mm overall	1·2·4 mix. 2% steel 490 mm square or 300 × 760 mm	2900 mm square 690 mm deep
8	10 × 10 in × 60 lb U.C. (260 × 256 mm × 89 kg) 370 × 370 mm overall	1·1½·3 mix. 2% steel 490 mm square or 300 × 760 mm	3130 mm square 840 mm deep
7	10 × 10 in × 89 lb U.C. (276 × 261 mm × 133 kg) 380 × 370 mm overall	1·1½·3 mix. 2% steel 540 mm square or 300 × 900 mm	3510 mm square 920 mm deep
6	10 × 10 in × 112 lb U.C. (289 × 265 mm × 167 kg) 400 × 370 mm overall	1·1½·3 mix. 3% steel 540 mm square or 300 × 900 mm	3740 mm square 1000 mm deep
5	10 × 10 in × 112 lb U.C. (289 × 265 mm × 167 kg) 400 × 370 mm overall	1·1½·3 mix. 3% steel 590 mm square or 300 × 1070 mm	3960 mm square 1070 mm deep
4	12 × 12 in × 133 lb U.C. (340 × 314 mm × 198 kg) 450 × 420 mm overall	1. 1½·3 mix. 3% steel 610 mm square or 300 × 1220 mm	4190 mm square 1150 mm deep
3	12 × 12 in × 133 lb U.C. (340 × 314 mm × 198 kg) 450 × 420 mm overall	1·1½·3 mix. 4% steel 610 mm square or 300 × 1220 mm	4420 mm square 1220 mm deep
2	12 × 12 in × 161 lb U.C. (353 × 318 mm × 240 kg) 460 × 430 mm overall	1·1½·3 mix. 4% steel 640 mm square or 300 × 1320 mm	4650 mm square 1320 mm deep
1st	12 × 12 in × 161 lb U.C. (353 × 318 mm × 240 kg) 460 × 430 mm overall	1·1½·3 mix. 4% steel 660 mm square or 300 × 1480 mm	4880 mm square 1370 mm deep

Example 2

see Figure 8.1.2.

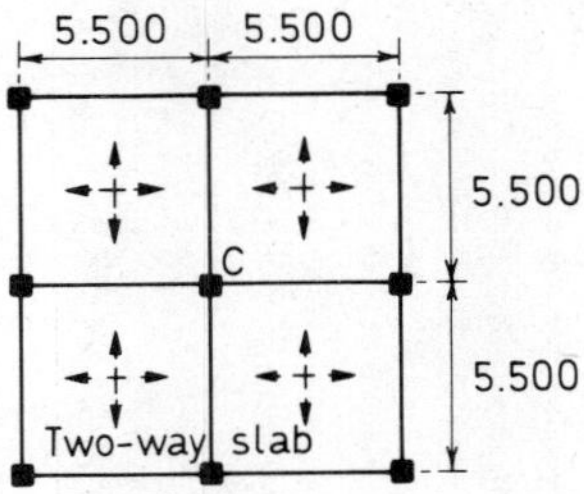

Figure 8.1.2

Slab thickness = 150 mm for roof and all floors.

Roof beam
Steel 12 × 6½ in × 31 lb U.B. (307 × 166 mm × 46 kg).
Reinforced concrete. Rib

= 230 × 150 mm or 230 × 200 mm overall
= 380 × 150 mm or 380 × 200 mm overall.

Floor beam
Steel 12 × 6½ in × 36 lb U.B. (311 × 167 mm × 54 kg).
Reinforced concrete, approximately as roof beams.

Example 2 Figure 8.1.2	Cased steel column (C) below	R.C. column (C) below	R.C. Foundation
Roof	6 × 6 in × 15·7 lb U.C. (152 × 152 mm × 23 kg) 250 × 250 mm overall	1 : 2 : 4 mix. 0·8% steel 200 × 200 mm	1070 mm square 300 mm deep
11th Floor	6 × 6 in × 15·7 lb U.C. (152 × 152 mm × 23 kg) 250 × 250 mm overall	1 : 2 : 4 mix. 1% steel 280 × 280 mm	1600 mm square 380 mm deep
10	6 × 6 in × 25 lb U.C. (162 × 154 mm × 37 kg) 270 × 260 mm overall	1 : 2 : 4 mix. 2% steel 320 mm square or 300 × 330 mm	1910 mm square 460 mm deep
9	8 × 8 in × 31 lb U.C. (203 × 203 mm × 46 kg) 300 × 300 mm overall	1 : 2 : 4 mix. 2% steel 370 mm square or 300 × 430 mm	2210 mm square 510 mm deep
8	8 × 8 in × 35 lb U.C. (206 × 204 mm × 52 kg) 310 × 310 mm overall	1 : 1½ : 3 mix. 2% steel 380 mm square or 300 × 450 mm	2440 mm square 610 mm deep
7	8 × 8 in × 48 lb U.C. (216 × 206 mm × 71 kg) 320 × 310 mm overall	1 : 1½ : 3 mix. 2% steel 410 mm square or 300 × 560 mm	2600 mm square 690 mm deep
6	8 × 8 in × 58 lb U.C. (222 × 209 mm × 86 kg) 330 × 320 mm overall	1 : 1½ : 3 mix. 3% steel 410 mm square or 300 × 560 mm	2820 mm square 750 mm deep
5	10 × 10 in × 60 lb U.C. (260 × 256 mm × 89 kg) 370 × 370 mm overall	1 : 1½ : 3 mix. 3% steel 450 mm square or 300 × 640 mm	3050 mm square 840 mm deep
4	10 × 10 in × 72 lb U.C. (267 × 258 mm × 107 kg) 370 × 370 mm overall	1 : 1½ : 3 mix. 3% steel 470 mm square or 300 × 760 mm	3200 mm square 920 mm deep
3	10 × 10 in × 89 lb U.C. (276 × 261 mm × 133 kg) 380 × 370 mm overall	1 : 1½ : 3 mix. 4% steel 470 mm square or 300 × 760 mm	3430 mm square 990 mm deep
2	10 × 10 in × 89 lb U.C. (276 × 261 mm × 133 kg) 380 × 370 mm overall	1 : 1½ : 3 mix. 4% steel 500 mm square or 300 × 790 mm	3580 mm square 990 mm deep
1	10 × 10 in × 112 lb U.C. (289 × 265 mm × 167 kg) 400 × 370 mm overall	1 : 1½ : 3 mix. 4% steel 510 mm square or 300 × 870 mm	3740 mm square 1070 mm deep

Example 3

See Figure 8.1.3

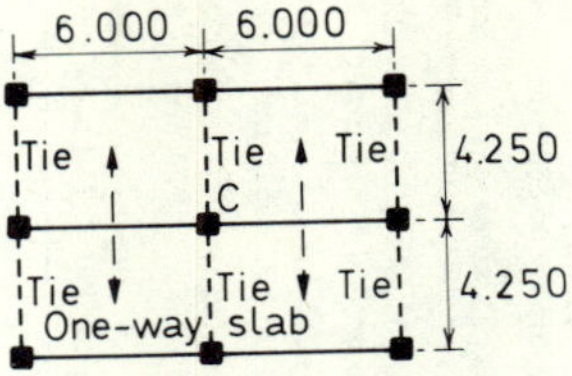

Figure 8.1.3

Slab thickness = 125 mm for roof and all floors

Roof beams
Steel. 14 × 6¾ in × 38 lb U.B. (359 × 172 mm × 57 kg).
Reinforced concrete. Rib = 250 × 150 mm i.e. 380 × 150 mm overall.
Steel tie beams. Say 8 × 5¼ in × 17 lb U.B. (203 × 133 mm × 25 kg).

Floor beams
Steel 14 × 6¾ in × 45 lb U.B. (364 × 173 mm × 67 kg)
Reinforced concrete, approximately as roof beams.
Steel tie beams as roof.

Example 3 Figure 8.1.3	Cased steel column (C) below	R.C. column (C) below	R.C. Foundation
Roof	6 × 6 in × 15·7 lb U.C. (152 × 152 mm × 23 kg) 250 × 250 mm overall	1 : 2 : 4 mix. 0·8% steel 200 mm square	990 mm square 300 mm deep
11th floor	6 × 6 in × 15·7 lb U.C. (152 × 152 mm × 23 kg) 250 × 250 mm overall	1 : 2 : 4 mix. % steel 280 mm square	1450 mm square 300 mm deep
10	6 × 6 in × 25 lb U.C. (162 × 154 mm × 37 kg) 270 × 260 mm overall	1 : 2 : 4 mix. 2% steel 290 mm square	1760 mm square 460 mm deep
9	8 × 8 in × 31 lb U.C. (203 × 203 mm × 46 kg) 300 × 300 mm overall	1 : 2 : 4 mix. 2% steel 330 mm square or 300 × 380 mm	2060 mm square 510 mm deep
8	8 × 8 in × 31 lb U.C. (203 × 203 mm × 46 kg) 300 × 300 mm overall	1 : 1½ : 3 mix. 2% steel 350 mm square or 300 × 380 mm	2290 mm square 610 mm deep
7	8 × 8 in × 35 lb U.C. (206 × 204 mm × 52 kg) 310 × 310 mm overall	1 : 1½ : 3 mix. 2% steel 370 mm square or 300 × 430 mm	2440 mm square 610 mm deep
6	8 × 8 in × 48 lb U.C. (216 × 206 mm × 71 kg) 320 × 310 mm overall	1 : 1½ : 3 mix. 3% steel 380 mm square or 300 × 450 mm	2590 mm square 690 mm deep
5	8 × 8 in × 58 lb U.C. (222 × 209 mm × 86 kg) 330 × 320 mm overall	1 : 1½ : 3 mix. 3% steel 410 mm square or 300 × 560 mm	2820 mm square 760 mm deep
4	8 × 8 in × 58 lb U.C. (222 × 209 mm × 86 kg) 330 × 320 mm overall	1 : 1½ : 3 mix. 3% steel 440 mm square or 300 × 610 mm	2980 mm square 840 mm deep
3	10 × 10 in × 60 lb U.C. (260 × 256 mm × 89 kg) 370 × 370 mm overall	1 : 1½ : 3 mix. 4% steel 440 mm square or 300 × 610 mm	3130 mm square 920 mm deep
2	10 × 10 in × 89 lb U.C. (267 × 258 mm × 107 kg) 370 × 370 mm overall	1 : 1½ : 3 mix. 4% steel 460 mm square or 300 × 690 mm	3280 mm square 920 mm deep
1	10 × 10 in × 89 lb U.C. (276 × 261 mm × 133 kg) 380 × 370 mm overall	1 : 1½ : 3 mix. 4% steel 470 mm square or 300 × 710 mm	3430 mm square 990 mm deep

8.2 GENERAL GUIDE

Long Span Plate and Lattice Girders

Depth of girder = about $\frac{1}{10}$ to $\frac{1}{14}$ of the span.

Lattice Shell Roofs in Steel

Spans up to 35 m are economic.

Lattice Steel Arches

One type of north light structure has concrete shells 75 mm thick spanning between lattice steel arches at 6 to 8 m centres. Clear spans as are from 25 to 65 m.

Double-Layer Space Grids (Steel)

Depth = about $\frac{1}{20}$ to $\frac{1}{30}$ of span.

Examples. Depth for spans about 30 m = about 1 m to 1·5 m.

Depth for a span of 100 m = about 3·5 m.

Reinforced Concrete Flat Plate Floors

Suitable for light and medium loadings, e.g. flats and offices.

Examples: Domestic loading and grid spacing of 4·5 m plate thickness = about 150 mm.

Office and classroom loading and grid spacing of 5 m plate thickness = about 200 mm.

Reinforced Concrete Flat Slab (Mushroom) Floor

Suitable for heavy distributed loads, say imposed (live) loads of 5 kN/m^2 or more as in warehouses.

Bays should be square or nearly square with a column grid of about 6–6·5 m. For these conditions the slab thickness will be from 230 to 300 mm.

Reinforced Concrete Rectangular or Diagonal Grid (Waffle) Floors

A depth-span ratio of $1/30$ is possible.

Shell Roofs with Long-Span Barrel Vaults

See Figure 8.2.1a.

Maximum economic span = about 30 to 45 m.

Width = about $1/2$ to $1/5$ of the span with maximum width of about 12 m.

Radius of curvature = about 6 to 6·5 m for small spans
= about 9 m for spans from about 15 to 30 m
= about 12 m for spans over 30 m.

Thickness of shell = about 60 to 75 mm.

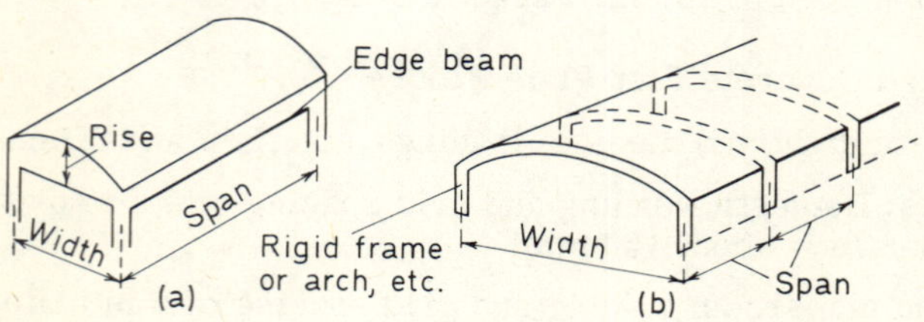

Figure 8.2.1

Shell Roofs with Short-Span Barrel Vaults

See Figure 8.2.1b

Frames = about 9-12 m apart.

Total rise = not less than $1/10$ span or chord width whichever is greater.

In-situ Reinforced Concrete Staircases

Stairs spanning horizontally between supporting walls or stringer beams (See Figure 8.2.2a)

For spans up to about 2m thickness of waist is about 80 mm.

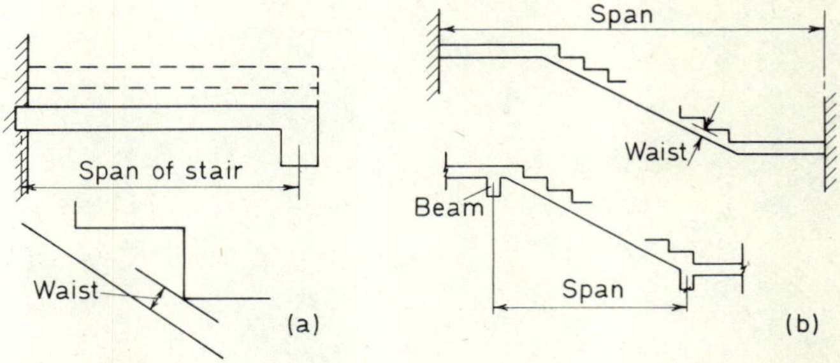

Figure 8.2.2

Stairs spanning longitudinally between supports at top and bottom of flight (no supports at sides, see Figure 8.2.2b)

Thickness of waist = approximately $1/_{25}$ of horizontal span.

Prestressed Concrete Floor Slabs

For spans up to about 10 m for flat roofs, flats and offices, minimum thickness of slab itself is about $1/_{40}$ of span.

Prestressed Concrete Beams

Depth = about $1/_{20}$ to $1/_{25}$ of span.

Timber Joists

The clear span of softwood joists is usually limited to 5 m.

At this span size of joists at 400 mm centres with domestic loading = 225 × 63 mm.

INDEX